# HOMEWORK HELPERS

# Pre-Calculus

## DENISE SZECSEI

D1051222

CAREER PRESS
Franklin Lakes NJ

HOMEWORK HELPERS: PRE-CALCULUS
EDITED BY KATHRYN HENCHES
TYPESET BY EILEEN DOW MUNSON
Cover design by Lu Rossman/Digi Dog Design NYC
Printed in the U.S.A. by Book-mart Press

To order this title, please call toll-free 1-800-CAREER-1 (NJ and Canada: 201-848-0310) to order using VISA or MasterCard, or for further information on books from Career Press.

The Career Press, Inc., 3 Tice Road, PO Box 687,
Franklin Lakes, NJ 07417
www.careerpress.com

## Library of Congress Cataloging-in-Publication Data

Szecsei, Denise.
    Homework helpers : Pre-calculus / by Denise Szecsei.
        p. cm.
    Includes bibliographical references and index.
    ISBN-13: 978-156414-940-4
    ISBN-10: 1-56414-940-4
        1. Functions—Problems, exercises, etc. 2. Mathematics. 3. Mathematics—Problems, exercises, etc. I. Title.

    QA331.S975 2007
    512--dc22

                                                        2006100141

# Acknowledgments

This book was a group effort, and I would like to thank the people who helped transform it from the electrons on my computer screen into the object you are holding in your hand.

I would like to thank Michael Pye, Kristen Parkes, and everyone else at Career Press who worked on this project. I appreciate the time and efforts of Jessica Faust, who was instrumental in making the connections that started things rolling.

Kendelyn Michaels helped in the development of this book, and Alic Szecsei continued to check my work and my typing. Even though he is taking a geometry class, he doesn't want his algebra skills to atrophy.

Thanks to my family for their help throughout the writing and editing stage. Their patience and understanding as yet another deadline approaches is appreciated.

# Contents

# Preface

## Welcome to *Homework Helpers: Pre-Calculus!*

The primary goal of a pre-calculus class is to prepare students for calculus. Many of the problems in calculus involve analyzing functions and solving equations. When learning algebra, students typically learn one problem-solving strategy at a time, but the complex problems that are solved in calculus often require using multiple problem-solving strategies. Pre-calculus is an introduction to how to successfully combine several techniques to solve a single problem.

You can think of pre-calculus as an opportunity to delve more deeply into familiar functions while practicing your algebraic skills. Whether we are talking about athletic skills, musical skills, or algebraic skills, they all have a tendency to get rusty when they aren't used. The more you know about linear, quadratic, polynomial, rational, exponential, and logarithmic functions, the easier it will be to understand the ideas being discussed in calculus.

Although the emphasis of pre-calculus is on preparing a student for calculus, there are many applications of pre-calculus. Modeling the population of a species using exponential functions, determining how long it takes for an investment to double in value, and calculating the dimensions of a dog run that maximize the enclosed area, are examples of problems that can be solved using the techniques discussed in pre-calculus. Sequences and series are important ideas that will be explored in calculus, and are instrumental in proving the Fundamental Theorem of Calculus.

Learning pre-calculus involves looking at problems from a new perspective. You will learn how to combine algebraic skills with analytical reasoning to deepen your familiarity with the basic functions that are discussed throughout calculus. An important aspect of mathematics is pattern recognition. Pre-calculus introduces you to the process of looking for patterns, making conjectures, and developing your mathematical intuition. It will also help to improve your reasoning skills.

I wrote this book with the hope that it will help anyone who is struggling to understand the ideas in pre-calculus or is just curious about the subject. Reading a math book can be a challenge, but I tried to use everyday language to explain the concepts being discussed. Looking at solutions to math problems can sometimes be confusing, so I tried to explain each of the steps I used to get from Point A to Point B. Keep in mind that learning mathematics is not a spectator sport. In this book I have worked out many examples, and I have supplied practice problems at the end of most of the lessons. Work these problems out on your own as they come up, and check your answers against the solutions at the end of each chapter. Aside from any typographical errors on my part, our answers should match.

I hope that in reading this book you will develop an appreciation for the subject of pre-calculus and the field of mathematics. In addition, I hope that it will help you avoid some of the bumps and bruises that students typically suffer when they study calculus.

# Algebra

Algebra can be thought of as a language of numbers. Numbers are the tools used to communicate mathematical ideas. This chapter will focus on revisiting the rules that must be followed in order to use numbers effectively.

## Lesson 1-1: Real Numbers

To develop the real numbers, we start with the **counting numbers**, which are the numbers 1, 2, 3, 4, ... (the three dots mean that the list of numbers is unending).

Zero was the last number to be discovered. It is a symbol used to denote "nothing." If you add zero to the collection of counting or natural numbers, you will have the whole numbers. The **whole numbers** are the numbers 0, 1, 2, 3, ....

Numbers can be used to count things that you have, and they can also count things that you owe. Negative integers are thought of as the *opposite* of the counting numbers. The **negative integers** are the numbers −1, −2, −3, −4, ....

The **integers** consist of the positive integers, the negative integers, and zero. They can be written ..., −3, −2, −1, 0, 1, 2, 3, .... Notice that the dots represent that the numbers go on forever in both directions.

A **rational number** is a number that can be written as the ratio of two integers. For example, the numbers $\frac{1}{2}$, $-\frac{2}{3}$, and $\frac{10}{3}$ are rational numbers.

A rational number is a number that can be written as $\frac{p}{q}$, where $p$ and $q$ are integers and $q$ is any number other than zero. The number $p$ is called the

**numerator** and $q$ is called the **denominator**. Notice that there's more than one way to represent a given rational number: $\frac{2}{3}$, $\frac{6}{9}$, and $\frac{-20}{-30}$ all represent the same rational number. The integer 2 is a rational number because $2 = \frac{2}{1}$. In fact, *every* integer is a rational number. Whenever you are working with whole numbers and fractions in the same problem, remember that a whole number can be thought of as a fraction whose denominator is 1.

All rational numbers also have a decimal representation that either terminates, or repeats. The numbers $\frac{3}{4} = 0.75$ and $\frac{3}{5} = 0.6$ have decimal representations that terminate. The numbers $\frac{1}{3} = 0.3333...$, $\frac{5}{11} = 0.454545...$, and $-\frac{131}{333} = -0.393393...$ have decimal representations that do not terminate, but repeat a pattern.

Not all decimal representations terminate or repeat. Numbers that are written as non-terminating and non-repeating decimals are called **irrational numbers**. An example of an irrational number is $\sqrt{2}$. An irrational number cannot be written as the ratio of two integers.

Taken together, the rational numbers and the irrational numbers form the set of **real numbers**. Breaking the real numbers into groups (the natural numbers, integers, rational numbers, irrational numbers, and real numbers), is one way to classify numbers in general.

Another way to classify a number is to look at the numbers that divide it evenly. When one number divides another number **evenly**, the remainder is 0. *Every* number is evenly divisible by 1, and *every* number is evenly divisible by itself. The number 1 is called the **trivial factor** (or divisor) because it divides evenly into every other number. The number 1 is the *only* trivial factor, because 1 is the only number that divides evenly into every other number. All other factors of a number are considered to be **non-trivial**.

A number is **even** if it is evenly divisible by 2. For example, 22 is even because when you divide 22 by 2 the remainder is 0. A number is **odd** if it is not evenly divisible by 2. For example, 15 is odd because when you divide 15 by 2 the remainder is 1 (not 0). Determining whether a number is even or odd is one example of a classification of numbers based on the numbers that divide them.

A number is **prime** if the only numbers that divide into it evenly are 1 and itself. For example, 7 is prime because the only numbers that divide into it, and leave a remainder of 0, are 1 and 7. A number is **composite** if there are numbers other than 1 and itself that divide into it evenly. The number 6 is composite because, besides 1 and 6, 2 divides into it evenly (as does 3). There is only one even number that is also a prime number: 2. All other even numbers are divisible by 2, and hence are composite. Odd numbers may be prime or composite. Examples of odd *prime* numbers are 3, 5, and 7; examples of odd *composite* numbers are 9 and 15. All natural numbers *greater than 1* can be classified as either prime or composite.

Besides looking at individual numbers on their own, we can also look at pairs of numbers. One way to classify a pair of numbers is according to the numbers that divide each number in the pair. A pair of numbers is **relatively prime** if the only common factor of the two numbers is 1. The numbers 15 and 8 are an example of a pair of numbers that are relatively prime. The only non-trivial factors of 15 are 3, 5, and 15, and the only non-trivial factors of 8 are 2, 4, and 8, so the only number that divides both 15 and 8 is 1. Notice that both 15 and 8 are *composite* numbers, when classified individually, yet they are *relatively prime* when classified as a pair. The numbers 7 and 21 are *not* relatively prime, because 7 is a non-trivial factor of both 7 and 21. So even though 7 is a prime number, it is *not* relatively prime to 21.

The **greatest common factor** of a pair of numbers is the largest number that evenly divides into both numbers. If two numbers are relatively prime then their greatest common factor is 1. Actually, their *only* common factor is 1. If two numbers are not relatively prime then their greatest common factor will be greater than 1. One way to find the greatest common factor of a pair of numbers is to first factor each number into its prime factors. Then match up, prime by prime, the prime numbers that appear in both factorizations. If any prime numbers repeat, you must take that into consideration. The greatest common factor of 54 and 72 is 18.

The **absolute value** of a real number represents the size, or *magnitude*, of that number. It can be interpreted as how far away from 0 the number is on the number line. The absolute value of a number $a$ is always positive, and is written symbolically as $|a|$. The absolute value of a positive number is itself. We can write this symbolically as:

$$\text{If } a > 0, \text{ then } |a| = a.$$

The absolute value of 0 is 0, and the absolute value of a negative number is the opposite of that number:

$$\text{If } a < 0, \text{ then } |a| = -a.$$

In general, if $a$ is any real number, then $|a| = |-a|$. We can write the absolute value of a number using the formula:

$$|a| = \begin{cases} -a & \text{if } a < 0 \\ a & \text{if } a \geq 0 \end{cases}.$$

You can read this formula as a fork in the road. The absolute value of a number depends on the sign of the number inside the absolute value symbols: $|a| = -a$ if $a < 0$ and $|a| = a$ if $a \geq 0$. In order to use this formula you must first check to see whether the contents of the absolute value symbols are positive or negative.

## Lesson 1-2: Properties of Real Numbers

There are several properties that all real numbers share. Because these are properties that *all* real numbers have, I will use the variables $a$, $b$, and $c$ to represent *any* real number.

▶ **Closure:** The first property that all real numbers have is **closure** under addition and multiplication. In other words, the set of real numbers has the property that if you take any two real numbers and add them (or multiply them), what you'll end up with is a real number. We say that the set of real numbers is **closed** under addition and multiplication.

▶ **Commutative property of addition and multiplication:** The order in which two numbers are added or multiplied doesn't matter: $a + b = b + a$ and $a \times b = b \times a$.

▶ **Associative property of addition and multiplication:** If you have a long list of numbers to add (or multiply), then you can group them in any order: $a + (b + c) = (a + b) + c$ and $a \times (b \times c) = (a \times b) \times c$.

The next two properties have more to do with the relationships between numbers.

▶ **Trichotomy property of real numbers:** When you compare two numbers $a$ and $b$, only one of three things can be true:

     1. $a < b$

     2. $a = b$

     3. $a > b$

This is known as the **trichotomy property** of real numbers. In fact, you can compare any two real numbers using these three relations.

▶ **Transitive property of equality:** If two numbers are both equal to a third number, then the two numbers are equal to each other. This property is called the **transitive property of equality**. It can be stated more generally in terms of variables: if $a = b$ and $b = c$, then $a = c$.

At first glance it may appear that all numbers are created equally, but there are actually two numbers that deserve special attention. Those two numbers are 0 and 1. Zero is the *only* number that you can add to any other number and have no effect. Zero is called the **additive identity**, and this property can be written in general using the equation $a + 0 = a$. It may not seem very important…after all, nothing is nothing. But quantifying nothing is not as trivial as you might think.

The number 1 is important for a similar reason; 1 is the **multiplicative identity**. It is the *only* number that you can multiply any other number by and have no effect. This idea is written in general using the equation $a \times 1 = a$. The numbers 0 and 1 will play a significant role in solving many algebra problems.

The additive identity plays a role in the development of subtraction. It turns out that for every real number $a$, there is a unique real number, called the **additive inverse** of $a$, and denoted $-a$, such that $a + (-a) = 0$. In other words, the additive inverse of $a$, or negative $a$, is the unique number that you *add* to $a$ to get 0 (the additive identity). Every real number has an additive inverse. Notice that 0 is its own additive inverse. Subtraction is then defined in terms of addition: $a - b = a + (-b)$.

You may have been told that a negative number times a negative number is a positive number. The reason for that stems from the fact that $-a$ is the opposite of $a$: $-a$ is the unique number that, when it is added to $a$, gives you 0. In other words, $a + (-a) = 0$. What would be the opposite of $-a$? Well, in keeping with our notation, it would be $-(-a)$. But wait a minute! The opposite of $-a$ is $a$. Now it appears that you have two opposites of $-a$: $a$ and $-(-a)$, but you can't have two different opposites of a number. Additive inverses are unique, which means that each number can only have one additive inverse. The only way for this to make sense is if $a$ and $-(-a)$ are the same thing: $-(-a) = a$. We can apply this result specifically to the number 1: $-(-1) = 1$. This is interpreted as meaning that a negative times a negative equals a positive.

It's time to turn our attention to multiplicative inverses. For any real number $a$ (except 0), there is a unique real number, called the **multiplicative inverse** and denoted $a^{-1}$, satisfying the equation $a \cdot a^{-1} = 1$. The multiplicative inverse of $a$ is the unique number that you *multiply* $a$ by to get 1 (the multiplicative identity). Notice that 0 (the additive identity) is the *only* real number that doesn't have a multiplicative inverse. The multiplicative inverse of a number is also called the **reciprocal** of that number. There are two common ways to represent the multiplicative inverse of $a$; it can be written as $a^{-1}$ or as $\frac{1}{a}$. Division can then be defined in terms of multiplication. If $b \neq 0$, then $a \div b = \frac{a}{b} = a\left(\frac{1}{b}\right) = a\left(b^{-1}\right)$. The reciprocal of $\frac{a}{b}$ is just $\frac{b}{a}$.

The fact that 0 is the only real number that doesn't have a reciprocal (or multiplicative inverse) is worth exploring in more detail. Whenever you multiply 0 and any number $a$, the result is always 0: $0 \times a = 0$. Remember that the reciprocal of a number is the number you multiply by in order to get 1. If 0 had a reciprocal, say $b$, then that would mean that the product of 0 and $b$ would be 1! But that can't happen, because the product of 0 and any number is 0. So 0 is stuck. It is precisely because 0 times any number equals 0, that 0 can't have a reciprocal. This fact translates into the idea that $\frac{1}{0}$ (which means the reciprocal of 0) is meaningless.

Another special property of 0 follows along a similar line of thought. If two numbers are multiplied together and the result is 0, then one thing is certain: at least one of the two numbers has to be 0. To write this idea using equations, if $a$ and $b$ are any two numbers satisfying $a \times b = 0$, then either $a = 0$ or $b = 0$. It is possible that both $a$ and $b$ are zero, because $0 \times 0$ is certainly 0. The key idea is that when you have to compare a product of numbers to something, the best number to compare the product to is 0! This is a very important property, and 0 is the *only* real number that has it.

The last property of the real numbers that I will discuss in this section has to do with how to combine addition and multiplication. It is called the **distributive property**. The distributive property states that multiplication **distributes** over addition and can be expressed using the formula: $a \times (b + c) = a \times b + a \times c$. You can think of $a$ as being distributed to both $b$ and $c$.

# Lesson 1-3: Exponents and Radicals

Exponents are a "shorthand" way to represent how many times a number is multiplied by itself. They are useful, in part, due to the fact that they can be used to represent numbers that are either very large, such as the number of grains of sand on the beach, or very small in magnitude, such as the size of an atom.

There are several important rules for combining exponents. These rules will be very useful when we discuss exponential functions.

When you multiply two numbers with the same base, you add the exponents. This is called the **product rule for exponents**:

$$a^m \times a^n = a^{m+n}$$

When you divide two numbers with the same base, you subtract the exponents. This is called the **quotient rule for exponents**:

$$\frac{a^m}{a^n} = a^{m-n}$$

When you raise a power to a power, you multiply the two exponents. This is called the **power of a power rule**:

$$\left(a^m\right)^n = a^{m \times n}$$

When you raise a product to a power, each term in the product must be raised to that power. This is called the **power of a product rule**:

$$(a \times b)^n = a^n \times b^n$$

When you raise a quotient to a power, each term in the quotient must be raised to that power. This is called the **power of a quotient rule**:

$$\left(\frac{a}{b}\right)^n = \frac{a^n}{b^n}$$

There are a few other rules for manipulating exponents. The first rule is that any non-zero number raised to the power 0 is 1: $a^0 = 1$. This observation actually comes from the quotient rule.

If $a \neq 0$, then $\frac{a^n}{a^n} = a^{n-n} = a^0$.

We can also relate negative exponents and reciprocals:

$$a^{-n} = \left(a^n\right)^{-1} = \left(a^{-1}\right)^n = \left(\tfrac{1}{a}\right)^n.$$

Familiarity with the rules for exponents will be extremely important in Chapter 5.

# Lesson 1-4: Algebraic Expressions

An **algebraic expression** is a statement that combines numbers and variables using any of the four operations you have learned about: addition, subtraction, multiplication, and division. For example, the expression $a + 2$ just means that you take the number $a$ (whatever that is) and add 2 to it. If someone told you that $a = 5$, then you would know that $a + 2$ would be 7 (which is $5+2$). In order to evaluate an expression for a particular value of the variables, just replace the variables with their particular values and then perform the calculation. Be sure to pay attention to the order of operations when you are performing the calculation.

### Example 1

Evaluate the following algebraic expressions when $a = -2$ and $b = 3$:

a. $\dfrac{-3a}{a+b}$

b. $\dfrac{a-b}{a+b}$

c. $(2a + b)(b - a)$

**Solutions:**

a. $\dfrac{-3a}{a+b} = \dfrac{(-3)(-2)}{-2+3} = \dfrac{6}{1} = 6$

b. $\dfrac{a-b}{a+b} = \dfrac{-2-3}{-2+3} = \dfrac{-5}{1} = -5$

c. $(2a + b)(b - a) = (2 \times (-2) + 3)(3 - (-2)) = (-4 + 3)(3 + 2)$
$$= (-1)(5) = -5$$

# Lesson 1-5: Equations

An equation is a statement that two expressions are equal. The concept of equality is like a balanced scale. Whatever you have on the left side of the equal sign exactly equals whatever you have on the right side. You are not allowed to tip the scales and favor one side over another side. Whatever you do to one side of the equation you must also do to the other side of the equation.

There are two algebraic properties of equality. These properties dictate what you are allowed to do with equations. The first algebraic property of equality is known as the **addition property of equality**. It states that if $a = b$, then $a + c = b + c$. We did not favor one side of the equality over the other side; the quantity $c$ was added to both sides so that the balance is maintained. The addition property of equality is important because it allows us to add (or subtract) the same number from both sides of an equation without changing the validity of that statement.

The second algebraic property of equality is known as the **multiplication property of equality**. It states that if $a = b$, then $a \cdot c = b \cdot c$. Again, one side is not favored over the other side; both sides are treated equally because both sides are being multiplied by $c$. When applying this property, it does not matter whether $c$ is a positive or a negative number. It also does not matter whether $c$ is greater than 1 or less than 1. You can multiply (or divide) both sides of an equation by the same number and not change the validity of the statement.

These properties are crucial in your success in solving algebraic equations. In general, when you want to solve a linear equation involving one variable, the first thing you need to do is gather all of the terms that involve the variable over to one side of the equation, and move all of the terms that don't involve the variable over to the other side of the equation. If, after combining all of the terms, the coefficient in front of the variable is a number other than 1, you will need to multiply both sides of the equation by the reciprocal of the coefficient in front of the variable. Once you have done that, you should have an explicit equation for what numerical value the variable has to be. The last step is to check your work (using the original problem statement) to make sure that your answer is correct. This last step is the one that is most often skipped, but it's one that you really should get into the habit of doing. By checking your work you will know whether your answer is correct or not. If it's not, you'll have to go back and start over.

## Example 1

Solve the equation $5x + 4 = 2x + 10$.

**Solution:** Gather the variables on one side of the equation and the numbers on the other side. Then solve the equation:

$$5x + 4 = 2x + 10$$

Subtract 4 from both sides $\quad 5x + 4 - 4 = 2x + 10 - 4$

Simplify $\quad 5x = 2x + 6$

Subtract $2x$ from both sides $\quad 5x - 2x = 2x + 6 - 2x$

Simplify $\quad 3x = 6$

Multiply both sides by $\frac{1}{3}$ $\qquad$ $\frac{1}{3} \cdot (3x) = \frac{1}{3} \cdot 6$

Simplify $\qquad$ $x = 2$

Finally, we'll check our answer.

If $x = 2$ then $5x + 4 = 5 \cdot 2 + 4 = 14$

## Example 2

Solve the equation $3x + 4(x - 1) = 10$.

**Solution:** First distribute the 4, then solve for $x$:

$$3x + 4(x - 1) = 10$$

Distribute the 4 $\qquad$ $3x + 4x - 4 = 10$

Simplify $\qquad$ $7x - 4 = 10$

Add 4 to both sides $\qquad$ $7x - 4 + 4 = 10 + 4$

Simplify $\qquad$ $7x = 14$

Divide both sides by 7 $\qquad$ $\dfrac{7x}{7} = \dfrac{14}{7}$

Simplify $\qquad$ $x = 2$

Finally, check your answer.

If $x = 2$ then $3 \cdot 2 + 4(2 - 1) = 6 + 4 \cdot 1 = 10$. So our answer is correct.

## Example 3

Solve the equation $10 = \frac{5}{3}(x + 2)$.

**Solution:** Rewrite the equation so that the variable is on the left.

You can either distribute the $\frac{5}{3}$ or multiply both sides of the equation by its reciprocal and solve for $x$:

$$10 = \frac{5}{3}(x + 2)$$

Interchange the sides of the equation $\qquad$ $\frac{5}{3}(x + 2) = 10$

Multiply both sides by $\frac{3}{5}$ $\qquad$ $\frac{3}{5} \cdot \frac{5}{3}(x+2) = \frac{3}{5} \cdot 10$

Simplify $\qquad$ $(x+2) = 6$

Subtract 2 from both sides $\qquad$ $x+2-2 = 6-2$

Simplify $\qquad$ $x = 4$

All that remains is to check our answer.

If $x = 4$ then $\frac{5}{3}(4+2) = \frac{5}{3} \cdot 6 = 10$, so our answer is correct.

---

We can also solve equations that involve absolute values. Remember that we have an equation for the absolute value of a number:

$$|a| = \begin{cases} -a & \text{if } a < 0 \\ a & \text{if } a \geq 0 \end{cases}.$$

The way you remove the absolute value symbols in an expression depends on what is inside of the absolute value symbols. If the contents of the absolute value symbols is positive, you can just drop the symbols. If the contents of the absolute value symbols is negative, then you are allowed to drop the symbols only after you put a negative sign in front of whatever was inside. If there are variables inside of the absolute value symbols, you won't know whether the contents are positive or negative, so you must explore both possibilities. In doing so, you will generate two equations when you remove the absolute value symbols.

When working with an equation that involves a variable inside of an absolute value symbol, the first thing you must do is isolate the absolute value part of the equation. It doesn't matter whether there are numbers or variables outside of the absolute value symbols; everything must be moved to the other side. Once you have done this, use the definition of absolute value to write two equations, depending on whether the contents of the absolute value symbols are positive or negative. Generate two equations that you will need to solve (and check your answers).

## Example 4

Solve the equation $|x + 5| = 3$.

**Solution:** Again, the absolute value symbol is already isolated. Generate the two equations and solve each one by subtracting 5 from both sides:

$$x + 5 = 3 \qquad\qquad -(x + 5) = 3$$
$$x + 5 - 5 = 3 - 5 \qquad\qquad x + 5 = -3$$
$$x = -2 \qquad\qquad x + 5 - 5 = -3 - 5$$
$$x = -8$$

So we have two solutions: $x = -2$ or $x = -8$. We must check each one.

Is $|-2 + 5| = |3|$ equal to 3? Yes.

Is $|-8 + 5| = |-3|$ equal to 3? Yes.

So both answers are correct: $x = -2$ or $x = -8$.

## Example 5

Solve the equation $|x + 2| + 3 = 6$.

**Solution:** We first need to isolate the absolute value by subtracting 3 from both sides. Then we will generate two equations that we can solve:

$$|x + 2| + 3 = 6$$
$$|x + 2| = 3$$

$$x + 2 = 3 \qquad\qquad -(x + 2) = 3$$
$$x + 2 - 2 = 3 - 2 \qquad\qquad x + 2 = -3$$
$$x = 1 \qquad\qquad x + 2 - 2 = -3 - 2$$
$$x = -5$$

The two solutions are $x = 1$ and $x = -5$. We need to check each solution.

Is $|1 + 2| + 3 = |3| + 3$ equal to 6? Yes.

Is $|-5 + 2| + 3 = |-3| + 3$ equal to 6? Yes.

So both solutions are correct: $x = 1$ or $x = -5$.

## Example 6

Solve the equation $2|x + 1| = 3$.

**Solution:** Again, we need to isolate the absolute value by dividing both sides of the equation by 2. Once we have done that we can generate the two equations and solve each one:

$$2|x+1| = 3$$

$$\frac{2|x+1|}{2} = \frac{3}{2}$$

$$|x+1| = \frac{3}{2}$$

$$x+1 = \frac{3}{2} \qquad\qquad -(x+1) = \frac{3}{2}$$

$$x+1-1 = \frac{3}{2}-1 \qquad\qquad x+1 = -\frac{3}{2}$$

$$x = \frac{1}{2} \qquad\qquad x+1-1 = -\frac{3}{2}-1$$

$$x = -\frac{5}{2}$$

Let's check each solution:

If $x = \frac{1}{2}$, we have $2\left|\frac{1}{2}+1\right| = 2\left|\frac{3}{2}\right| = 2\cdot\frac{3}{2} = 3$,

and if $x = -\frac{5}{2}$ we have $2\left|-\frac{5}{2}+1\right| = 2\left|-\frac{3}{2}\right| = 2\cdot\frac{3}{2} = 3$,

so both of our answers work: $x = \frac{1}{2}$ or $x = -\frac{5}{2}$.

_____

## Example 7

Solve the equation $|2x + 1| = 5$.

**Solution:** The absolute value is already isolated, so we just have to generate our two equations, solve them, and check our solutions:

$$|2x + 1| = 5 \qquad\qquad -(2x + 1) = 5$$

$$2x = 4 \qquad\qquad\qquad 2x + 1 = -5$$

$$x = 2 \qquad\qquad\qquad\quad 2x = -6$$

$$x = -3$$

Now to check our solution.

If $x = 2$ then $|2 \cdot 2 + 1| = |5| = 5$

and if $x = -3$ then $|2 \cdot (-3) + 1| = |-5| = 5$.

Both solutions check out: $x = 2$ or $x = -3$.

## Example 8

Solve the equation $2|x + 2| + 2 = 4$.

**Solution:** First isolate the absolute value:

$$2|x + 2| + 2 = 4$$
$$2|x + 2| = 2$$
$$|x + 2| = 1$$

Next, generate our two equations and solve them:

$$x + 2 = 1 \qquad\qquad -(x + 2) = 1$$
$$x = -1 \qquad\qquad x + 2 = -1$$
$$\qquad\qquad\qquad x = -3$$

Finally, check our answers. If $x = -1$ then $2|-1 + 2| + 2 = 2|1| + 2 = 4$ and if $x = -3$ then $2|-3 + 2| + 2 = 2|-1| + 2 = 4$. Both solutions work, so $x = -1$ or $x = -3$.

## Lesson 1-5 Review

Solve the following equations:

1. $7x + 5 = 2x - 10$

2. $4x + 2(x - 1) = 7$

3. $7x - 2(x - 4) = -2$

4. $12 = \frac{4}{5}(x - 2)$

5. $|x - 1| + 4 = 6$

6. $3|x - 1| = 5$

7. $|3x - 1| = 5$

8. $\frac{1}{2}|x + 3| + 1 = 3$

# Lesson 1-6: Inequalities

An **inequality** is an algebraic statement that compares two algebraic expressions that may not be equal. There are four basic inequalities shown in the chart on page 25.

A solution to an inequality is the collection of all numbers that produce a true statement when substituted for the variable in the inequality. In order to solve inequalities, we will need to talk about the two algebraic properties of inequalities. The algebraic properties of inequalities are very similar to the algebraic properties of equality.

| Symbol | Meaning | Example |
|:---:|:---|:---|
| $<$ | less than | $3 < 8$ |
| $\leq$ | less than or equal to | $3 \leq 8, 8 \leq 8$ |
| $>$ | greater than | $8 > 3$ |
| $\geq$ | greater than or equal to | $8 \geq 3, 8 \geq 8$ |

There is an **addition property of inequality**: if $a > b$, then $a + c > b + c$. The addition property of inequality holds regardless of the value of $c$: $c$ can be positive or negative, greater than 1 or less than 1. The addition property of inequality holds regardless of the value of $c$. There is also a **multiplication property of inequalities**, but you need to be careful when you apply it. If $a > b$ and $c > 0$, then $a \cdot c > b \cdot c$. This property requires that the number that you are multiplying by must be positive. This requirement is very important. If you multiply an inequality by a negative number, you must also remember to flip the inequality from $<$ to $>$ or vice versa. We can write this mathematically as:

$$\text{If } a < b \text{ and } c < 0, \text{ then } a \cdot c > b \cdot c.$$

This flipping rule also holds when you divide both sides of an inequality by a negative number.

Solving inequalities involves the same steps that solving equalities does: isolate the variable to one side of the inequality. The final answer should be written in interval notation.

## Example 1

Solve the inequality $2x + 10 \geq 7(x + 1)$.

**Solution:** First distribute the 7 on the right, then collect all the terms involving variables on one side, and the numbers on the other. Finally, solve for $x$:

$$2x + 10 \geq 7(x + 1)$$

Distribute the 7          $2x + 10 \geq 7x + 7$

Subtract 10 from both sides          $2x + 10 - 10 \geq 7x + 7 - 10$

Simplify          $-5x \geq -3$

Divide both sides by $-5$ and flip the inequality

$$\frac{-5x}{-5} \le \frac{-3}{-5}$$

Simplify

$$x \le \frac{3}{5}$$

The solution is all real numbers less than or equal to $\frac{3}{5}$, or $\left(-\infty, \frac{3}{5}\right]$.

## Example 2

Solve the inequality $12 > -2x - 6$.

**Solution:** Add 6 to both sides and then divide both sides by $-2$. Remember to flip the inequality:

$$12 > -2x - 6$$

Add 6 to both sides     $12 + 6 > -2x - 6 + 6$

Simplify     $18 > -2x$

Divide both sides by $-2$ and flip the inequality

$$\frac{18}{-2} < \frac{-2x}{-2}$$

Simplify     $-9 < x$

The solution is all real numbers greater than $-9$, or $(-9, \infty)$.

## Example 3

Solve the inequality: $10 \ge \frac{5}{3}(x+2)$

**Solution:** There are several ways to start this problem. You could distribute the $\frac{5}{3}$ and then move the terms around, or you could first multiply by $\frac{3}{5}$ and then simplify.

I will work the problem out the second way, and leave it to you to work it out the first way:

$$10 \geq \frac{5}{3}(x+2)$$

| | |
|---|---|
| Multiply both sides by $\frac{3}{5}$ | $\frac{3}{5} \cdot 10 \geq \frac{3}{5} \cdot \frac{5}{3}(x+2)$ |
| Simplify | $6 \geq x + 2$ |
| Subtract 2 from both sides | $6 - 2 \geq x + 2 - 2$ |
| Simplify | $4 \geq x$ |

The solution is all real numbers less than or equal to 4, or $(-\infty, 4]$.

To solve inequalities that involve absolute values, follow the same procedure that was used to solve equalities that involve absolute values. When you mix inequalities and absolute value symbols, you will end up creating two *inequalities* when you remove the absolute value symbols in the equation. And satisfying two inequalities in the same problem means that the solution will be a compound inequality.

Your approach to solving inequalities with absolute value should be almost identical to the approach you used to solve equations that involved absolute value. First, isolate the absolute value part of the inequality. Then create two inequalities and solve each one. Put the solutions to the two inequalities together to get the final solution.

Usually, when your equation involves an absolute value of something that is less than (or less than or equal to) a number, you will need to satisfy both of the inequalities that you generate, and your answer will consist of the regions where the two rays overlap. If your equation involves an absolute value of something that is greater than (or greater than or equal to) a number, your solution will be all of the numbers that lie on one ray or the other.

The inequality $|ax + b| > c$ is used to create two inequalities, depending on whether $ax + b$ is positive or negative. If $ax + b$ is positive, then $|ax + b| = ax + b$ and the inequality $|ax + b| > c$ becomes $ax + b > c$. If $ax + b$ is negative, then $|ax + b| = -(ax + b)$ and the inequality $|ax + b| > c$ becomes $-(ax + b) > c$. The two inequalities $ax + b > c$ and $-(ax + b) > c$ represent the *initial meaning* of the inequality $|ax + b| > c$. If we multiply the inequality $-(ax + b) > c$ by $-1$, we get $ax + b < -c$. The two inequalities $ax + b > c$ or $ax + b < -c$ represent the *simplified meaning* of the inequality $|ax + b| > c$.

The inequalities $ax + b > c$ and $ax + b < -c$ represent two rays that do not overlap. The solution to the inequality $|ax + b| > c$ will be the set of points that satisfy one inequality or the other.

This same analysis is done for the other three types of inequalities you will encounter in this book, and the results are shown in the table below. When I work out problems I will remove the absolute values using the initial meaning, and then carefully transform the inequality, flipping the inequality when necessary.

| Inequality | Initial Meaning | Simplified Meaning |
|---|---|---|
| $|ax + b| > c$ | $ax + b > c$ or $-(ax + b) > c$ | $ax + b > c$ or $ax + b < -c$ |
| $|ax + b| \geq c$ | $ax + b \geq c$ or $-(ax + b) \geq c$ | $ax + b \geq c$ or $ax + b \leq -c$ |
| $|ax + b| < c$ | $ax + b < c$ and $-(ax + b) < c$ | $ax + b < c$ and $ax + b > -c$ |
| $|ax + b| \leq c$ | $ax + b \leq c$ and $-(ax + b) \leq c$ | $ax + b \leq c$ and $ax + b \geq -c$ |

## Example 4

Solve the inequality $|x + 2| < 4$.

**Solution:** The absolute value is already isolated. So all we need to do is generate our two inequalities by substituting in for the absolute value symbols. We then solve both inequalities:

$$x + 2 < 4 \qquad -(x + 2) < 4$$
$$x < 2 \qquad x + 2 > -4$$
$$\qquad \qquad x > -6$$

The solution is the set of real numbers that are less than 2 and greater than −6: (−6, 2).

## Example 5

Solve the inequality $|2x - 1| + 2 \geq 7$.

**Solution:** First isolate the absolute value by subtracting 2 from both sides:

$$|2x - 1| + 2 \geq 7$$
$$|2x - 1| \geq 5$$

Next, generate the two inequalities and solve them both:

$$|2x - 1| \geq 5 \qquad -(2x - 1) \geq 5$$
$$2x \geq 6 \qquad\qquad 2x - 1 \leq -5$$
$$x \geq 3 \qquad\qquad 2x \leq -4$$
$$\qquad\qquad\qquad x \leq -2$$

The solution is the set of real numbers that are either greater than or equal to 3, or less than or equal to −2, or $(-\infty, -2] \cup [3, \infty)$.

## Lesson 1-6 Review

Solve the following inequalities and write your answers using interval notation.

1. $|x - 3| \geq 3$

2. $|3x + 2| \leq 8$

3. $|2x + 3| - 5 < 7$

4. $2|3x - 9| > 10$

## Answer Key

### Lesson 1-5

1. $x = -3$

2. $x = \frac{3}{2}$

3. $x = -2$

4. $x = 17$

5. $x = 3 \text{ or } x = -1$

6. $x = \frac{8}{3} \text{ or } x = -\frac{2}{3}$

7. $x = 2 \text{ or } x = -\frac{4}{3}$

8. $x = 1 \text{ or } x = -7$

### Lesson 1-6

1. $(-\infty, 0] \cup [6, \infty)$

2. $\left[-\frac{10}{3}, 2\right]$

3. $\left(-\frac{15}{2}, \frac{9}{2}\right)$

4. $\left(-\infty, \frac{4}{3}\right) \cup \left(\frac{14}{3}, \infty\right)$

# Functions

The concept of a function is very important in mathematics. Functions can be used to describe, or model, many situations in our everyday lives. In economics, functions can be used to calculate income tax, interest earned from an investment, and monthly loan payments. In science, functions can be used to predict the pressure exerted by a gas, the energy released in a chemical reaction, and the occurrence of the next lunar eclipse. The study of pre-calculus involves analyzing some basic elementary functions. As you will soon discover, solving problems in pre-calculus is very algebraic in nature, so it will be important to have a strong foundation in algebra.

## Lesson 2-1: Representing Functions

A **function** is a set of instructions that establishes a relationship between two quantities. A function has input and output values. The input is called the **domain** and the output is called the **range**. The variable used to describe the elements in the domain is called the **independent variable**. The variable used to describe the output is called the **dependent variable**, because it *depends* on the input. An important feature of a function is that every input value has only one corresponding output value.

A function is given a name. Sometimes a function is given the name $y$, other times it is given the name $f(x)$. Keep in mind that the *name* of the function is not as important as what the function *does*.

A function can be represented in a variety of ways. Functions can be described using words, a formula, a table, or a graph.

Describing a function using words is often involved in word problems. For example, suppose that my cell phone plan costs $39.99 per month plus $0.35 per minute for air time over 500 minutes. From this description, I could reason out what my cell phone bill would be if I used 600 minutes in one month.

Using a formula to define a function is a convenient way to describe the function in mathematical terms instead of using words. For example, the function that takes the input and squares it and then adds 5 could be described by the formula $f(x) = x^2 + 5$. The order of operations is very important when working with formulas. Exponentiation comes before addition, so when you read $x^2 + 5$ you should realize that you need to square $x$ first, then add 5. When you see a function such as $f(x) = x^2 + 5$ you should recognize the important features: the variable that appears in parentheses in the function name is the independent variable, and is sometimes referred to as the **argument** of the function. The formula for the function shows you how to transform the input into the output. This formula instructs you to take whatever is in parentheses and square it first, then add 5. Technically, it doesn't really matter what is in parentheses. For the function $f(x) = x^2 + 5$, we can see that

$$f(2) = 2^2 + 5 = 4 + 5 = 9$$

and $f(4) = 4^2 + 5 = 16 + 5 = 21,$

but in a similar manner, $f(\Theta) = \Theta^2 + 5,$

and $f(x + 1) = (x + 1)^2 + 5 = (x^2 + 2x + 1) + 5 = x^2 + 2x + 6.$

As long as you follow the rules and replace the independent variable with the object in parentheses, nothing can go wrong.

Scientists often collect data from various instruments and record this information in a table. These tables can represent functions, and provide an easy way to describe a complex, or unknown, formula. Reading a table is straightforward. If a table is written vertically, then the column on the left represents the independent variable, or the input, and the column on the right represents the dependent variable, or the output. For example, the function described in the table shown here does not have a formula associated with it, but by reading the table it is possible to determine that $f(3) = 7$ and $f(9) = 12$.

| $x$ | $f(x)$ |
|-----|--------|
| 3   | 7      |
| 9   | 12     |

If a table is written horizontally, then usually the top row represents the independent variable and the bottom row represents the dependent variable. For example, the function defined by the table shown here also does not have a formula associated with it, but you can see that $g(0) = -5$ and $g(1) = 2$.

| $x$ | 0 | 1 |
|-----|----|----|
| $g(x)$ | −5 | 2 |

The graph of function is usually presented using the Cartesian coordinate system. The Cartesian coordinate system is named for Rene Descartes, who is credited with inventing this system. We use this system, which is sometimes called the coordinate plane, to locate points and draw figures.

In the Cartesian coordinate system, we use two real number lines, one horizontal and one vertical, and let them intersect at 0. The point of intersection is called the **origin**. The horizontal number line is called the **x-axis** and it is used to record the values of the input, or independent variable. The vertical number line is called the **y-axis** and it is used to record the function values, or the output. Taken together, the x-axis and the y-axis are called the **coordinate axes**.

Two numbers are used to describe the location of a point in the plane, and they are recorded in the form of an ordered pair $(x, y)$, where the first number represents the horizontal distance (using a horizontal number line) from the y-axis to the point, and the second number represents the vertical distance (using a vertical number line) from the x-axis to the point. The first coordinate of the ordered pair is called the **x-coordinate** and the second coordinate is called the **y-coordinate**. Points that are on the x-axis have a y-coordinate equal to 0, and points that are on the y-axis have their x-coordinate equal to 0. Points that lie to the right of the y-axis have a positive x-coordinate, and points to the left of the y-axis have a negative x-coordinate. Similarly, points above the x-axis have a positive y-coordinate, and points below the x-axis have a negative y-coordinate.

The coordinate axes divide the plane into four parts, called Quadrants. Quadrant I consists of those points that have positive values for both their x-coordinate and y-coordinate. Quadrant I is located in the upper right part of the plane. We continue on to Quadrants II, III, and IV in a counterclockwise progression. Quadrant II consists of those points that have a negative value for their x-coordinate and a positive value for their y-coordinate. Quadrant III consists of those points that have negative values for both their x-coordinate and their y-coordinate. Finally, Quadrant IV consists of those points that have a positive value for their x-coordinate and a negative value for their y-coordinate. The Cartesian coordinate system is shown in Figure 2.1 on page 34.

The signs for the $x$-coordinates and $y$-coordinates are summarized in the table below.

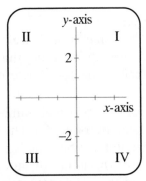

| Quadrant | Coordinate Signs for $(x, y)$ |
|----------|-------------------------------|
| I | $(+, +)$ |
| II | $(-, +)$ |
| III | $(-, -)$ |
| IV | $(+, -)$ |

*Figure 2.1*

A function can be thought of as a collection of ordered pairs $(a, b)$, where $a$ represents the value of the independent variable, or input, and $b$ represents the corresponding value of the dependent variable, or output. The graph of a function is then the graph of these ordered pairs in the coordinate plane.

If the graph of the function crosses the $x$-axis, then we say that the function has an $x$-intercept. The **$x$-intercept** of a function is the value of the $x$-coordinate of the point where the function crosses the $x$-axis. Because any point on the $x$-axis has a $y$-coordinate of 0, the $x$-intercept of a function corresponds to a point of the form $(a, 0)$, where $a$ represents the value of the input that results in an *output* of 0. The $x$-intercepts of a function are often called the **roots** of the function. Not all functions have $x$-intercepts.

If the graph of the function crosses the $y$-axis, then we say that the function has a $y$-intercept. The **$y$-intercept** of a function is the value of the $y$-coordinate of the point where the function crosses the $y$-axis. Because any point on the $y$-axis has an $x$-coordinate of 0, the $y$-intercept of a function corresponds to a point of the form $(0, b)$, where $b$ represents the value of the output that results from an *input* of 0. Not all functions have $y$-intercepts. In order for a function to have a $y$-intercept, 0 must be in the domain of the function.

# Lesson 2-2: The Domain of a Function

The **domain** of a function represents the allowed values of the independent variable. If a function is described using words, then the domain needs to incorporate the context of the description of the function. For example, if a function describes the number of buses needed for a field trip as a function of the number of passengers expected, then the domain of this function cannot include any negative numbers. It would make no sense to transport –50 passengers!

The description of a function using a formula may or may not include a domain. If the domain is not indicated, then it is safe to assume that the domain is the set of all real numbers that, when substituted in for the independent variable, produce real values for the dependent variable. To find the domain of a function, I recommend starting with the set of all real numbers and whittling down the list.

There are two things that are frowned upon in the mathematical community. The first thing that is forbidden is to divide a non-zero number by 0; quotients such as $\frac{3}{0}$ are meaningless. The second thing that is not allowed in the world of real numbers is to take an even root of a negative number; there is no real number that corresponds to $\sqrt{-4}$. To find the domain of a function that involves an even root (a square root, a fourth root, and so on) set whatever is under the radical to be greater than or equal to 0 and find the solutions to the inequality. Then toss out the problem points that give a 0 in the denominator.

### Example 1

Find the domain of the function $f = \{(1, 2), (3, -2), (4, 8)\}$.

**Solution:** The domain of this function consists of the values that the independent variable, or the first coordinate, takes on: $\{1, 3, 4\}$.

### Example 2

Find the domain of the function $f(x) = \sqrt{x+4}$.

**Solution:** Set the expression under the radical to be greater than or equal to 0 and solve:

$$x + 4 \geq 0$$

$$x \geq -4$$

Now check to see if any values of $x$ result in a denominator of 0. Because the radical is in the numerator of the expression, the denominator is an implied 1. There are no values of $x$ that will turn 1 into 0, so the domain of the function is $x \geq -4$. We can write the domain of this function using interval notation: $[-4, \infty)$. The bracket next to $-4$ means that $-4$ is included in the domain. A bracket can be used when your domain involves an inequality of the form $\geq$ or $\leq$. We use the symbol $\infty$ to represent the concept of infinity. The parentheses next to $\infty$ acknowledges that our independent variable can never actually reach infinity, but it can take on values as large as we can imagine.

## Example 3

Find the domain of the function $f(x) = \dfrac{4}{x+2}$.

**Solution:** Because there are no radicals involved in the function, start with the set of real numbers and toss out any values of the variable that will make the denominator equal to 0.

The denominator of the fraction $\dfrac{4}{x+2}$ is $x + 2$, so we must make sure that $x + 2 \neq 0$, or $x \neq -2$.

The domain of the function is all real numbers other than $-2$. We can write that symbolically as $x \neq -2$. In interval notation, the domain is $(-\infty, -2) \cup (-2, \infty)$. The parentheses by $-2$ mean that $-2$ is not included in the domain. We use the symbol $\cup$ to represent the union of the two intervals, and the domain consists of the points that are in either $(-\infty, -2)$ or $(-2, \infty)$.

## Example 4

Find the domain of the function $f(x) = \dfrac{\sqrt{2-x}}{x+5}$.

**Solution:** Start with the radical and then deal with the denominator. Set the contents of the radical to be greater than or equal to 0 and solve the inequality:

$$2 - x \geq 0$$

$$x \leq 2$$

Now focus on the denominator of the function: $x + 5 \neq 0$, which means that $x \neq -5$. The domain of the function is the set of all real numbers less than or equal to 2, excluding −5. We can write this using interval notation: $(-\infty, -5) \cup (-5, 2]$. Notice that −5 has parentheses next to it, to indicate that $x \neq -5$, and that 2 has a bracket next to it, because the inequality $x \leq 2$ means that 2 is included in the domain.

---

## Lesson 2-2 Review

Find the domain of the following functions:

1. $f(x) = \sqrt{6 - x}$      2. $f(x) = \dfrac{x}{\sqrt{x+1}}$      3. $f(x) = \dfrac{\sqrt{x+5}}{x-2}$

# Lesson 2-3: Operations on Functions

The *algebra* of real numbers establishes rules for how to combine real numbers. Numbers can be added, subtracted, multiplied, and divided. Algebraic expressions are an abstract way to represent numbers, so it is only natural that we are able to add, subtract, multiply, and divide algebraic expressions as well. There is one additional thing that we can do with functions: We can take their composition.

Think of a function as a transformation of things from the domain, or the input, to things in the range, or the output. If a function $f$ has a domain $X$ and it's range is a subset of a set $Y$, then we use the notation $f: X \rightarrow Y$ to represent the idea that $f$ is a function from $X$ to $Y$, or $f$ maps objects in $X$ to objects in $Y$. Suppose that $g$ is a function whose domain is $Y$ and whose range is a subset of a set $Z$. Then we can use the functions $f$ and $g$ to define a new function whose domain is $X$ and whose range is contained in $Z$. This new function would take an element in $X$ to its corresponding element in $Y$ using the function $f$, and then take that element in $Y$ to an element in $Z$ using the function $g$. This process of stringing functions from set to set is called the **composition** of functions. Because $f$ takes things from $X$ to $Y$, and $g$ takes things from $Y$ to $Z$, the new function "$g$ composed with $f$" takes things *directly* from $X$ to $Z$.

We write the composition of $f$ and $g$ in this order described as $g \circ f$. The functions are applied *right* to *left*: $g \circ f(x)$ means *first* apply the function $f$ to $x$, and *then* apply the function $g$ to the result. We can write $g \circ f(x) = g(f(x)); g \circ f(x)$ is read "$g$ composed with $f$," or "$g$ of $f(x)$."

The order in which we compose things matters. In general,
$g \circ f(x) \neq f \circ g(x)$.
In other words, $g(f(x)) \neq f(g(x))$. We can look at a complicated function
like $h(x) = \sqrt{3x+1}$ in terms of the composition of two functions. If we
define $f(x) = 3x + 1$ and $g(x) = \sqrt{x}$, then $h = g \circ f$. Remember that functions
are just instructions for what to do with whatever is in parentheses. The
function $g(x) = \sqrt{x}$ instructs us to take whatever is in parentheses and
put it under a radical. The function $f(x) = 3x + 1$ instructs us to take
whatever is in parentheses and triple it and then add 1. So

$$g(f(x)) = \sqrt{f(x)} = \sqrt{3x+1}.$$

Alternatively, we could substitute in for $f(x)$ using its formula and then
apply $g$: $g(f(x)) = g(3x + 1) = \sqrt{3x+1}$.
Either way you evaluate $g \circ f(x)$, you get the function $h(x)$.

Now let's look at the composition in reverse order. Notice that, in
this situation,

$$f \circ g(x) = f(g(x)) = 3g(x) + 1 = 3\sqrt{x} + 1.$$

Alternatively, if we first substitute in for $g(x)$ using its formula and then
apply $f$, we have:

$$f \circ g(x) = f(g(x)) = f(\sqrt{x}) = 3\sqrt{x} + 1.$$

This illustrates the fact that the order in which you compose functions
matters. In the example we just looked at, $g \circ f(x) = \sqrt{3x+1}$, and

$$f \circ g(x) = 3\sqrt{x} + 1.$$

In general, $f \circ g(x)$ is a different function than $g \circ f(x)$.

To find the composition of functions that are defined using a table,
just evaluate the composition one step at a time.

## Example 1

If $f(x)$ and $g(x)$ are defined by the table:

| x | f(x) | g(x) |
|---|------|------|
| 0 | 2 | 4 |
| 1 | 3 | 2 |
| 2 | 6 | 0 |
| 3 | 8 | 1 |

find the following:

a. $f(3)$      b. $(f \circ g)(1)$      c. $(f \circ f)(1)$      d. $(f \circ g \circ g)(3)$

**Solution:** Write each composition using function notation and evaluate the composition from the inside out:

a. $f(3) = 8$

b. $(f \circ g)(1) = f(g(1)) = f(2) = 6$

c. $(f \circ f)(1) = f(f(1)) = f(3) = 8$

d. $(f \circ g \circ g)(3) = f(g(g(3))) = f(g(1)) = f(2) = 6$

---

## Lesson 2-3 Review

Find $g \circ f(x)$ and $f \circ g(x)$ for the following pairs of functions:

1. $f(x) = \sqrt{6-x}$, $g(x) = 2x+1$

2. $f(x) = \dfrac{1}{3x+6}$, $g(x) = \dfrac{1}{3}(x-6)$

3. $f(x) = x^2 + 2$, $g(x) = \sqrt{x-2}$

4. If $f(x)$ and $g(x)$ are defined by the table:

| $x$ | $f(x)$ | $g(x)$ |
|---|---|---|
| 0 | 1 | 3 |
| 1 | 3 | 2 |
| 2 | 0 | 1 |
| 3 | 4 | 0 |

find the following:

a. $g(3)$      b. $(f \circ g)(1)$      c. $(f \circ f)(1)$      d. $(g \circ f \circ g)(3)$

# Lesson 2-4: Transformations of Functions

The graph of a function can be moved around the coordinate plane. The process by which a graph is moved is referred to as a **transformation** of a function. In general, the transformation of the graph of a function can involve a **shift** (also called a translation), a **reflection**, a **stretch,** or a **contraction**. Transformations can occur vertically, meaning that the result is a change in the value of the $y$-coordinate, or horizontally, meaning that they result in a change in the $x$-coordinate of the graph. We will consider each of these transformations individually.

To shift a function *vertically*, add a constant to the dependent variable, or to the function. For example, to shift the function $f(x) = x^2$ *up* 3 units, simply *add* 3 to $f(x)$: the graph of the function $g(x)$, defined by $g(x) = f(x) + 3 = x^2 + 3$, is just the graph of the original function $f(x)$ shifted up 3 units. To shift the function $f(x) = x^2$ *down* 5 units, simply *subtract* 5 from (or add −5 to) $f(x)$ to create a new function $h(x)$: $h(x) = f(x) - 5 = x^2 - 5$. The graphs of $f(x)$, $g(x)$, and $h(x)$ are shown in Figure 2.2.

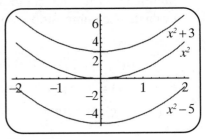

To shift a function *horizontally*, add a constant to the independent variable, *Figure 2.2* or the *argument* of the function. To shift the function $f(x) = x^2$ to the *left* 3 units, simply *add* 3 to the argument of $f(x)$ to create a new function $g(x)$: $g(x) = f(x + 3) = (x + 3)^2$. To shift the function $f(x) = x^2$ to the *right* 5 units, simply subtract 5 from (or add −5 to) the argument of $f(x)$ to create a new function $h(x)$: $h(x) = f(x - 5) = (x - 5)^2$. The graphs of $f(x)$, $g(x)$, and $h(x)$ are shown in Figure 2.3.

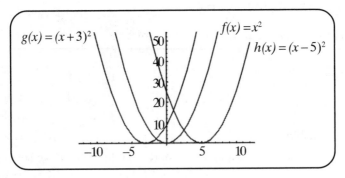

*Figure 2.3*

This may help you remember the difference between a vertical shift and a horizontal shift: a *vertical* shift moves the graph up or down, and the up/down scale is measured along the y-axis. A vertical shift involves adding a constant to $y$, which is what is meant by the notation $f(x) + c$. A *horizontal* shift moves the graph left or right, and the left/right scale is measured along the x-axis. So a horizontal shift involves adding a constant to $x$, or every place where $x$ appears in the formula, which is what is meant by the notation $f(x + c)$.

Reflecting a graph is like looking at it in a mirror. Every point on one side of the mirror is sent to the corresponding point on the opposite side of the mirror. To reflect the graph of a function across the *x*-axis, the *x*-axis acts as a mirror, and every point *above* the *x*-axis is sent to the corresponding point *below* the *x*-axis; every point *below* the *x*-axis is sent to the corresponding point *above* the *x*-axis. In this process, the signs of the *y*-coordinates of every point of the function must be changed: positive *y*-coordinates become negative and negative *y*-coordinates become positive. The way to change the sign of a number is to multiply by –1, and that is what we do to reflect the graph of a function across the *x*-axis: multiply the function by –1. The graph of the function $g(x) = -f(x)$ is a reflection of the graph of $f(x)$ across the *x*-axis. For example, if $f(x) = x^2$, then the graph of $g(x) = -f(x) = -x^2$ is the reflection of the graph of $f(x)$ across the *x*-axis, as shown in Figure 2.4.

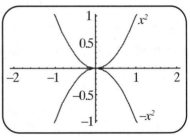

*Figure 2.4*

To reflect the graph of a function across the *y*-axis, the signs of the *x*-coordinates of every point of the function must be changed. In order to do this, the sign of the *argument* of the function must change. The graph of the function $g(x) = f(-x)$ is a reflection of the graph of $f(x)$ across the *y*-axis. For example, if $f(x) = x + 1$, then the graph of $g(x) = f(-x) = -x + 1$ is the reflection of the graph of $f(x)$ across the *y*-axis, as shown in Figure 2.5.

The effect of stretching a graph is to "draw it out" so that it has the same shape but occupies more space. Contracting a graph effectively shrinks the graph, or makes it narrower. We can stretch or contract a graph vertically or horizontally. To *stretch* the graph of a function vertically, multiply each *y*-coordinate of the function by a constant that is greater than 1. To *contract* a function vertically, multiply each *y*-coordinate of a function by a positive constant that is less than 1. For example, the function $f(x) = x^2$ can be stretched vertically by a factor of 3 by multiplying the function by 3: $g(x) = 3f(x) = 3x^2$.

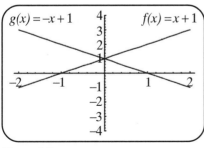

*Figure 2.5*

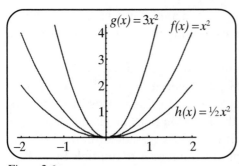

Figure 2.6

The function $f(x) = x^2$ can be contracted vertically by a factor of 2 by multiplying the function by $\frac{1}{2}$: $h(x) = \frac{1}{2}f(x) = \frac{1}{2}x^2$.

Both of these examples are shown in Figure 2.6.

To stretch the graph of a function horizontally, multiply the argument of the function by a positive constant that is less than 1. To contract the graph of a function, multiply the argument of the function by a constant that is greater than 1. For example, the function $f(x) = x^2$ can be stretched horizontally by a factor of 2 by multiplying the argument of the function by $\frac{1}{2}$:

$$g(x) = f\left(\tfrac{1}{2}x\right) = \left(\tfrac{1}{2}x\right)^2 = \tfrac{1}{4}x^2.$$

The function $f(x) = x^2$ can be contracted horizontally by a factor of 3 by multiplying the function by 3:
$$h(x) = f(3x) = (3x)^2 = 9x^2.$$

Both of these examples are shown in Figure 2.7.

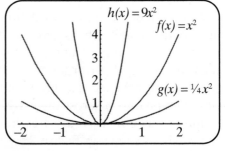

Figure 2.7

Notice that in the previous examples of stretching and contracting the function $f(x) = x^2$, stretching vertically and contracting horizontally seemed to have the same overall effect. In general, that will *not* be true.

## Lesson 2-4 Review

Transform the following functions:

1. Shift $f(x) = x^3$ to the right 4 units.

2. Shift $f(x) = 3x - 1$ up 8 units.

3. Shift $f(x) = \sqrt{x}$ to the left 2 units.

4. Shift $f(x) = \dfrac{1}{x}$ down 5 units.

5. Stretch $f(x) = \sqrt{x}$ vertically by a factor of 3.

6. Contract $f(x) = x^3$ horizontally by a factor of $\frac{1}{4}$.

# Lesson 2-5: Even and Odd Functions

The graph of a function can have some inherent symmetry. There are two important symmetries that we will discuss in this lesson. The first type of symmetry is when the $y$-axis serves as a mirror. The second type of symmetry has to do with the arrangement of the points of a function relative to the origin.

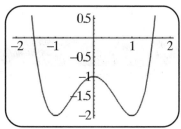

In the graph of the function shown in Figure 2.8, the $y$-axis acts as a mirror. Every point to the left of the $y$-axis has a corresponding point to the right.

The $x$-coordinates of the point on the

*Figure 2.8*

left and its corresponding point on the right have the same magnitude; only their signs are different. The $y$-coordinates of the point on the left and its corresponding point on the right are exactly the same. Reflecting the graph of this function across the $y$-axis will not change the function. Recall that the transformation of reflecting across the $y$-axis can be written as $f(-x)$. Symmetry with respect to reflection across the $y$-axis can be stated algebraically as: $f(-x) = f(x)$. If a function is symmetric with respect to the $y$-axis, that function is called an **even** function.

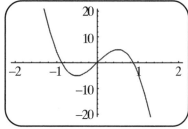

In the graph of the function shown in Figure 2.9, the graph of the function is symmetric with respect to the origin. Notice that points in Quadrant I have corresponding points in Quadrant III and points in Quadrant II have corresponding points in Quadrant IV. This can be seen as reflecting a point across *both* the $x$-axis and the $y$-axis.

*Figure 2.9*

Reflecting across the $y$-axis can be written as $f(-x)$, and reflecting across the $x$-axis can be written as $-f(x)$. Combining these transformations results in the overall transformation $-f(-x)$. The only point that doesn't change when you reflect it across both axes is the origin, which is why we refer to this type of symmetry as being symmetric with respect to the origin. A function that is symmetric with respect to the origin must satisfy the equation $f(x) = -f(-x)$. This is often rewritten as: $f(-x) = -f(x)$. Functions that are symmetric with respect to the origin are called **odd** functions.

Functions that are symmetric are easier to graph. If you pick your points carefully, you can get away with plotting half as many points and seeing the whole picture. To graph an even function, find the $y$-intercept and then evaluate the function for some positive values of $x$. Using the symmetry of the function you can then graph the corresponding points to the left of the $y$-axis.

## Example 1

Fill in the table for the following *even* function:

| $x$ | $f(x)$ |
|-----|--------|
| $-2$ | 5 |
| $-1$ | 2 |
| 0 | 1 |
| 1 | |
| 2 | |

**Solution:** Because the function is even, $f(-x) = f(x)$, so we have:

| $x$ | $f(x)$ |
|-----|--------|
| $-2$ | 5 |
| $-1$ | 2 |
| 0 | 1 |
| 1 | 2 |
| 2 | 5 |

To graph an odd function, first realize that every odd function must pass through the origin. This can be seen by evaluating the equation $f(-x) = -f(x)$ when $x = 0$:

$f(-x) = -f(x)$
$f(-0) = -f(0)$
$f(0) = -f(0)$

The only number that is unchanged when you multiply by –1 is 0, so $f(0) = 0$ and we see that an odd function must pass through the origin. To graph an odd function, evaluate the function for some positive values of $x$. Using the symmetry of the function and the fact that the graph of the function must pass through the origin, you can graph the corresponding points.

## Example 2

Fill in the table for the following odd function:

| $x$ | $f(x)$ |
| --- | --- |
| –2 | 5 |
| –1 | 2 |
| 0 | |
| 1 | |
| 2 | |

**Solution:** Because the function is odd, $f(-x) = -f(x)$ and the function must pass through the origin. The completed table is:

| $x$ | $f(x)$ |
| --- | --- |
| –2 | 5 |
| –1 | 2 |
| 0 | 0 |
| 1 | –2 |
| 2 | –5 |

## Lesson 2-6: Average Rate of Change

The average rate of change of a function over an interval is defined as the ratio of the change in the function over the interval divided by the length of the interval. It is a measure of how much the function (or independent variable) changes relative to how much the independent variable changes. The average rate of change of $f(x)$ over the interval $[a, b]$ is defined as the ratio:

$$\text{average rate of change} = \frac{f(b) - f(a)}{b - a}.$$

This ratio should look familiar: the average rate of change of $f(x)$ over the interval $[a, b]$ is just the slope of the line passing through the points $(a, f(a))$ and $(b, f(b))$. The phrase "average rate of change" should make you think of the slope of a line.

## Example 1

Find the average rate of change of $f(x) = x^2$ over the interval $[1, 3]$.

**Solution:** Evaluate the function at the two endpoints to generate two points. Then find the slope of the line that passes through the two points: $f(3) = 3^2 = 9$, so the first point is $(3, 9)$. To find the second point, evaluate $f(1): f(1) = 1^2 = 1$. The second point is $(1, 1)$. The average rate of change of $f(x) = x^2$ over the interval $[1, 3]$ is:

$$\text{average rate of change} = \frac{f(3) - f(1)}{3 - 1} = \frac{9 - 1}{2} = \frac{8}{2} = 4 .$$

The average rate of change can be used to determine your average speed on a long trip or to estimate how fast a population is changing over time.

## Example 2

The population of Florida was 16.0 million in 2000 and 17.8 million in 2005. What was the average rate of change of the population over that time interval?

**Solution:** The average rate of change of the population is the ratio of the change in population to the change in time:

$$\text{average rate of change} = \frac{17.8 - 16.0}{2005 - 2000} = \frac{1.8}{5} = .36$$

The average rate of change of the population was 0.36 million people per year, or 360,000 people per year.

## Lesson 2-6 Review

1. Find the average rate of change of the function $f(x) = \sqrt{2x + 1}$ over the interval $[0, 4]$.

2. A house was purchased for $107,500 in 2000 and sold for $250,000 in 2006. Find the average rate of change of the price of the house over that time interval.

# Lesson 2-7: Difference Quotients

The difference quotient is very important in calculus. It forms the basis of the definition of the derivative, which is discussed extensively in calculus. The difference quotient is an abstract way of talking about slopes and average rates of change.

We have practiced evaluating functions for specific values of the independent variable. We have practiced evaluating functions when the argument is replaced by an algebraic expression. We have learned how to create new functions by adding, subtracting, multiplying, and dividing old functions. As you will soon see, evaluating the difference quotient of a function involves all of these things.

As I mentioned earlier, the difference quotient is related to the slope of a line. This is the geometric significance of the difference quotient. Our focus will be on the algebraic aspects of the difference quotient...we will evaluate and simplify the difference quotient for a few elementary functions.

Given any function $f(x)$, the **difference quotient** of the function at a point $x$ is defined to be the ratio $\frac{f(x+h)-f(x)}{h}$. The difference quotient of a function $f(x)$ is actually a function of two variables: $x$ and $h$. Don't let the two variables bother you. Each variable represents a number, and as we will see later on, we will usually want $h$ to be a number that is small in magnitude. Looking at the difference quotient expression $\frac{f(x+h)-f(x)}{h}$, we can see right away that there will be problems if $h$ is ever equal to 0. In fact, one of our requirements is that $h \neq 0$. One of the games in calculus is to let $h$ be as small as possible, without actually becoming 0. We usually approach difference quotients in three steps:

Step 1. Evaluate $f(x+h)$.

Step 2. Evaluate $f(x+h) - f(x)$.

Step 3. Put the pieces together and evaluate $\frac{f(x+h)-f(x)}{h}$, simplifying the expression as much as possible.

## Example 1

Let $f(x) = 3x + 4$. Find the difference quotient $\dfrac{f(x+h)-f(x)}{h}$.

**Solution:**

Step 1: Evaluate this function when the argument is $x + h$:

$f(x) = 3x + 4$

$f(x + h) = 3(x + h) + 4 = 3x + 3h + 4$

Step 2: Evaluate $f(x + h) - f(x)$:

$f(x + h) - f(x) = (3x + 3h + 4) - (3x + 4)$

$f(x + h) - f(x) = 3x + 3h + 4 - 3x - 4$

$f(x + h) - f(x) = 3h$

Step 3: Evaluate the difference quotient. In Step 2 we simplified the numerator, and we can substitute the simplified form of

$f(x + h) - f(x)$ directly into the numerator of

$\dfrac{f(x+h)-f(x)}{h}$ :

$\dfrac{f(x+h)-f(x)}{h} = \dfrac{3h}{h} = 3$, as long as $h \neq 0$.

Let me emphasize that we are allowed to cancel the $h$s in the difference quotient as long as we stipulate that $h \neq 0$.

## Example 2

Let $f(x) = x^2 + 3x$. Evaluate the difference quotient $\dfrac{f(x+h)-f(x)}{h}$.

**Solution:**

Step 1: Evaluate this function when the argument is $x + h$:

$f(x) = x^2 + 3x$

$f(x + h) = (x + h)^2 + 3(x + h) = x^2 + 2xh + h^2 + 3x + 3h$

Step 2: Evaluate $f(x + h) - f(x)$:

$$f(x + h) - f(x) = (x^2 + 2xh + h^2 + 3x + 3h) - (x^2 + 3x)$$
$$f(x + h) - f(x) = x^2 + 2xh + h^2 + 3x + 3h - x^2 - 3x$$
$$f(x + h) - f(x) = 2xh + h^2\ 3h$$

Step 3: Evaluate the difference quotient. In Step 2 we simplified the numerator, and we can substitute the simplified form of $f(x + h) - f(x)$ directly into the numerator of

$$\frac{f(x+h)-f(x)}{h}:$$

$$\frac{f(x+h)-f(x)}{h} = \frac{2xh+h^2+3h}{h} = \frac{\cancel{h}(2x+h+3)}{\cancel{h}} = 2x+h+3,$$

as long as $h \neq 0$.

_____

This is an important idea in calculus, and it is in your best interest to practice working with the difference quotient and become familiar with the algebraic steps involved in simplifying it. The time you spend practicing now will be time well spent.

## Lesson 2-7 Review

Evaluate the difference quotient for the following functions:

1.  $f(x) = 6x - 5$    .

2.  $f(x) = 2x^2 - x$

## Answer Key

### Lesson 2-2 Review

1.  The domain of $f(x) = \sqrt{6 - x}$ is $x \leq 6$.
    In interval notation, the domain is $(-\infty, 6]$.

2.  The domain of $f(x) = \dfrac{x}{\sqrt{x+1}}$ is $x > -1$.
    In interval notation, the domain is $(-1, \infty)$.

3.  The domain of $f(x) = \dfrac{\sqrt{x+5}}{x-2}$ is $x \geq -5$ and $x \neq 2$.
    In interval notation, the domain is $[-5, 2) \cup (2, \infty)$.

## Lesson 2-3 Review

1. $g \circ f(x) = g(f(x)) = 2f(x) + 1 = 2\sqrt{6-1} + 1,$

   $f \circ g(x) = f(g(x)) = \sqrt{6 - g(x)} = \sqrt{6 - (2x+1)} = \sqrt{5 - 2x}$

2. $g \circ f(x) = g(f(x)) = \dfrac{1}{3}\left(\dfrac{1}{3x+6} - 6\right) = \dfrac{1}{3}\left(\dfrac{1}{3x+6} - \dfrac{6(3x+6)}{(3x+6)}\right) = \dfrac{1}{3}\left(\dfrac{-18x - 35}{3x+6}\right),$

   $f \circ g(x) = f(g(x)) = \dfrac{1}{3g(x)+6} = \dfrac{1}{3(\frac{1}{3}(x-6))+6} = \dfrac{1}{(x-6)+6} = \dfrac{1}{x}$

3. $g \circ f(x) = g(f(x)) = \sqrt{f(x) - 2} = \sqrt{(x^2 + 2) - 2} = \sqrt{x^2} = x$

   $f \circ g(x) = f(g(x)) = [g(x)]^2 + 2 = \left(\sqrt{x-2}\right)^2 + 2 = (x-2) + 2 = |x|$

4. a. $g(3) = 0$
   b. $(f \circ g)(1) = f(g(1)) = f(2) = 0$
   c. $(f \circ f)(1) = f(f(1)) = f(3) = 4$
   d. $(g \circ f \circ g)(3) = g(f(g(3))) = g(f(0)) = g(1) = 2$

## Lesson 2-4 Review

1. $f(x-4) = (x-4)^3$

2. $f(x) + 8 = (3x - 1) + 8 = 3x + 7$

3. $f(x+2) = \sqrt{x+2}$

4. $f(x) - 5 = \dfrac{1}{x} - 5$

5. $3f(x) = 3\sqrt{x}$

6. $f(4x) = (4x)^3$

## Lesson 2-6 Review

1. $\dfrac{f(4) - f(0)}{4 - 0} = \dfrac{3 - 1}{4} = \dfrac{1}{2}$

2. $\dfrac{250,000 - 107,500}{2006 - 2000} = \dfrac{142,500}{6} = 23,750$

## Lesson 2-7 Review

1. $f(x) = 6x - 5, f(x+h) = 6(x+h) - 5,$

   $\dfrac{f(x+h) - f(x)}{h} = \dfrac{(6(x+h) - 5) - (6x - 5)}{h} = \dfrac{6x + 6h - 5 - 6x + 5}{h} = \dfrac{6h}{h} = 6$

   as long as $h \neq 0$.

2. $f(x) = 2x^2 - x,$

$$f(x) = 2(x+h)^2 - (x+h) = 2(x^2 + 2xh + h^2) - x - h$$
$$= 2x^2 + 4xh + 2h^2 - x - h$$

$$\frac{f(x+h) - f(x)}{h} = \frac{(2x^2 + 4xh + 2h^2 - x - h) - (2x^2 - x)}{h} = \frac{4xh + 2h^2 - h}{h}$$

$$= \frac{\cancel{h}(4x + h - 1)}{\cancel{h}} = 4x + h - 1$$

as long as $h \neq 0$.

# Elementary Functions and Their Graphs

The ability to analyze functions is an important skill to develop. Linear functions, quadratic functions, and polynomials are examples of elementary functions. We will analyze these functions in detail, and then we will apply similar techniques to analyze more complex functions. These elementary functions appear throughout mathematics, and familiarity with them will make your future mathematical explorations much more meaningful.

## Lesson 3-1: Linear Functions

We will begin our study of elementary functions by talking about linear functions. **Linear** functions are functions of the form $f(x) = mx + b$, where m is called the **slope** and $b$ is the **y-intercept** of the function. For linear functions, m ≠ 0. For the special case where m = 0, the function $f(x) = b$ is called a **constant** function. A constant function is *not* a linear function.

The term "linear function" suggests a connection between these functions and lines. In fact, the graph of a linear function is a line that is neither vertical nor horizontal. The graph of any linear function is a line, but be careful: not all lines are categorized as linear functions. In particular, horizontal and vertical lines are not *linear* functions. A horizontal line is a *constant* function, and a vertical line is not a function at all. It may seem as though we are being needlessly particular, but this really is an important point. There are statements that we will make about linear functions that are not true for lines in general. For example, if two linear functions are perpendicular, then the product of their slopes is −1. This is not true for lines in general. A horizontal and a vertical line are perpendicular, but the slope of a vertical line is undefined, and the slope of a horizontal line is 0; we cannot talk about the product of something that is undefined and 0. It's alright to make the association between a linear function and a line, but you have to keep in mind that when we talk about linear functions, we are specifically ruling out vertical and horizontal lines.

The domain of a linear function is the set of all real numbers. A linear function is completely determined by two pieces of information: its slope and its y-intercept.

The equation of a linear function can be determined from any two points that lie on the line. If $(x_1, y_1)$ and $(x_2, y_2)$ are two points that lie on a line, we can calculate the slope of the line using the formula:

$$\text{slope} = \frac{y_2 - y_1}{x_2 - x_1}$$

The slope of the line connecting the two points can be regarded as the ratio of the *change* in the y-coordinates to the *change* in the x-coordinates. This is also commonly referred to as the *rise* of the line $(y_2 - y_1)$ divided by the *run* of the line $(x_2 - x_1)$, or "rise over run." If we use the symbol $\Delta$ to represent change, then we can write the formula for the slope as:

$$\text{slope} = m = \frac{y_2 - y_1}{x_2 - x_1} = \frac{\Delta y}{\Delta x},$$

where $\Delta y = y_2 - y_1$ and $\Delta x = x_2 - x_1$. As long as $\Delta x \neq 0$ and $\Delta y \neq 0$, the line will not be vertical or horizontal.

Once you know the slope of the line, you can use the point-slope formula to find the equation of the line: $y - y_1 = m(x - x_1)$. Remember that you started with two points, and you can use *either* of the two points in the point-slope formula: $y - y_1 = m(x - x_1)$ or $y - y_2 = m(x - x_2)$. Regardless of which point you use in the point-slope formula, when you simplify it and write the equation of the line in slope-intercept form, you should get the same overall equation.

The slope-intercept form of a line is: $y = mx + b$

Replacing $y$ with $f(x)$ yields the familiar linear function $f(x) = mx + b$, where m is the slope and b is the y-intercept. In other words, finding a linear function that passes through two points is the same thing as finding the equation of the line that passes through the two points and writing it in slope-intercept form.

## Example 1

Find the equation of the line that passes through the points $(2, -3)$ and $(4, 5)$.

**Solution:** First, find the slope of the line:

$$\text{slope} = \frac{y_2 - y_1}{x_2 - x_1} = \frac{-3 - 5}{2 - 4} = \frac{-8}{-2} = 4.$$

Then, use the point-slope formula and simplify. You can use either of the two points. I will use the point $(4, 5)$: $\quad y - y_1 = m(x - x_1)$

$$y - 5 = 4(x - 4)$$
$$y - 5 = 4x - 16$$
$$y = 4x - 11$$

To graph any linear function, all you need is two points. Any two points will do, but my favorite two points are the $x$-intercept and the $y$-intercept. Remember that you find the $x$-intercept by setting $y = 0$ and solving for $x$. To find the $y$-intercept of a function, let $x = 0$ and find $y$. Once you find two points (whether they are the intercepts or not), graph the two points in the Cartesian coordinate system and use a straightedge (or a ruler) to draw the line that passes through them.

## Example 2

Graph the linear function $f(x) = -2x + 4$.

**Solution:** Find two points, either by finding the two intercepts or evaluating the function for two different values of $x$. I will find the two intercepts. The $y$-intercept can be quickly determined from the function: it is the point $(0, 4)$. To find the $x$-intercept, set $f(x)$ equal to 0 and solve for $x$:

$$f(x) = -2x + 4 = 0$$
$$-2x = -4$$
$$x = 2.$$

The $x$-intercept is the point $(2, 0)$. Now we can graph the two intercepts and draw the graph of the function, as shown in Figure 3.1.

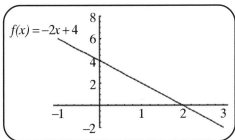

Figure 3.1

To find the y-intercept of a linear function, or *any* function for that matter, simply evaluate the function at $x = 0$. Every *linear* function will have exactly one y-intercept, and the y-intercept can be read off just by looking at the formula for the linear function: if $f(x) = mx + b$, the y-intercept is the point $(0, b)$. In general, if $x = 0$ is in the domain of a function, that function will have one y-intercept.

On the other hand, finding the x-intercepts of a linear function, or *any* function for that matter, involves setting the function equal to 0 and solving for *x*. All *linear* functions have exactly one x-intercept, but not all functions have x-intercepts, and some functions have more than one x-intercept. Finding the x-intercepts of a function is an important skill, and it goes by many names. Finding the roots of a function and finding the zeros of a function are two ways to rephrase the concept of finding the x-intercepts of a function.

Linear functions are useful in mathematical modeling. They are fairly easy to evaluate, and are uniquely determined using only two points. They can be used to extrapolate outside of a data set, and they can be used to interpolate between points in a data set.

## Example 3

A minor league baseball team plays in a park with a seating capacity of 10,000 spectators. With the ticket price set at $10, the average attendance at recent games has been 7,000 spectators. A market survey indicates that for each dollar the ticket price is lowered, the average attendance increases by 800 spectators. Find a linear function that models the attendance as a function of price, and find the price that will lead to a full house.

**Solution:** We need to express the attendance as a linear function of price. Attendance represents the dependent variable, and price represents the independent variable. We need to use variables for attendance and price. Let *a* represent attendance and *p* represent price. We are looking for the function $a(p)$. To find the equation of a line, we need two points. We are given one data point, corresponding to the point $(10, 7{,}000)$, and instructions on how to find a second point. If the price is lowered by $1 (meaning that the price is now $9), the attendance will be 7,800. The point $(9, 7{,}800)$ is another data point. With two points we can find the equation of the line that passes through them, but we first need to find the slope:

$$\text{slope} = \frac{\Delta a}{\Delta p} = \frac{7,800 - 7,000}{9 - 10} = -800$$

The fact that the slope is −800 should not be a surprise: a *decrease* in price of $1 results in an *increase* in attendance of 800. Now that we know the slope and a point that the line passes through, we can find the equation of the line. Remember that we can use either point to find the equation of a line; I will use the point (10, 7,000):

$$a - 7,000 = -800(p - 10)$$

$$a = -800p + 15,000$$

If the house is full, then there will be 10,000 spectators in attendance. We can use our model to solve for the price:

$$a = -800p + 15,000$$

$$10,000 = -800p + 15,000$$

$$-5,000 = -800p$$

$$p = 6.25$$

Charging a price of $6.25 will fill the house, according to our model.

---

It is important to realize the limitations of our mathematical models. There are many variables besides price that will affect attendance at a baseball game, but introducing these other variables will result in a much more complicated model. Linear models are a good place to start, despite their simplistic nature.

## Lesson 3-1 Review

1. Find the linear function that passes through the points (2, 5) and (5, 7).

2. Find the intercepts and graph the linear function $f(x) = 3x - 4$.

3. The cost for printing T-shirts for a local nonprofit organization includes $50 for the T-shirt design and $4 per shirt. Express the cost of printing the T-shirts as a linear function of the number of T-shirts that are printed, and determine the cost of printing 500 T-shirts.

# Lesson 3-2: Quadratic Functions

A quadratic function is a function of the form $f(x) = ax^2 + bx + c$, where $a$, $b$, and $c$ are constants, and $a \neq 0$. The function $f(x) = 3x^2 + 2x + 1$

is one example of a quadratic function. The constant $a$ is called the **leading coefficient** of the quadratic function, and $b$ is called the **linear coefficient**. The graph of a quadratic function is called a **parabola**. The domain of a quadratic function is the set of all real numbers. There are five important features of a quadratic function, and analyzing these features will help us to easily graph these types of functions.

The first feature of a quadratic function $f(x) = ax^2 + bx + c$ is that its graph will either "open" up or down, depending on the sign of the leading coefficient. If $a > 0$ the parabola will open up, or look like it bends upward, and if $a < 0$ the parabola will open down, or look like it bends downward. We refer to the direction of a curve as its **concavity**. A parabola that opens upward is called **concave up**, and a parabola that opens downward is called **concave down**. The graph of the quadratic function $f(x) = 2x^2 - x + 1$ is a parabola that opens upward, or is concave up, and the graph of the quadratic function $f(x) = -3x^2 + x - 7$ is a parabola that opens downward, or is concave down. Figure 3.2 shows the graphs of two parabolas: one that opens up and another that opens down.

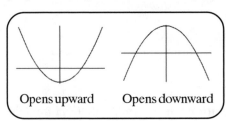

Opens upward　　Opens downward

*Figure 3.2*

The second feature of a quadratic function $f(x) = ax^2 + bx + c$ has to do with the symmetry of its graph. Every quadratic function has a vertical line that splits it into two symmetrical halves. This vertical line is called the **axis of symmetry**. The axis of symmetry of a parabola only depends on the leading coefficient and the linear coefficient of the quadratic function. The equation for the axis of symmetry is:

$$x = -\frac{b}{2a}$$

The quadratic function $f(x) = 2x^2 - 12x + 3$ opens upward and its axis of symmetry is:

$$x = -\frac{b}{2a} = -\frac{(-12)}{(2)(2)} = -\frac{(-12)}{4} = 3 \, .$$

The quadratic function $f(x) = -3x^2 - 12x + 5$ opens downward and its axis of symmetry is:

$$x = -\frac{b}{2a} = -\frac{(-12)}{(2)(-3)} = -\frac{(-12)}{(-6)} = -2 \, .$$

The third feature of a quadratic function has to do with the point where the axis of symmetry intersects the graph of the parabola. This point is called the **vertex** of the parabola. If the parabola opens up, then the vertex corresponds to the *lowest* point on the parabola. The lowest point of the graph corresponds to the **absolute minimum** value that the function achieves on its domain. If the parabola opens down, then the vertex corresponds to the *highest* point on the parabola. The highest point of the graph corresponds to the **absolute maximum** value that the function achieves on its domain. Because the vertex of a parabola is where the axis of symmetry intersects the function, the $x$-coordinate of the vertex and the value of the axis of symmetry are the same: $x = -\frac{b}{2a}$. The $y$-coordinate of the vertex is found by evaluating the function at $x = -\frac{b}{2a}$. In other words, the vertex of a parabola is the point $\left(-\frac{b}{2a}, f\left(-\frac{b}{2a}\right)\right)$.

## Example 1

Discuss the first three features of the quadratic function
$f(x) = 2x^2 + 8x - 1$.

**Solution:** The leading coefficient is positive, so the parabola is concave up. To find the axis of symmetry, evaluate

$$x = -\frac{b}{2a} : x = -\frac{8}{(2)(2)} = -2.$$

The axis of symmetry is the vertical line $x = -2$. The vertex is found by evaluating the function $f(x) = 2x^2 + 8x - 1$ at $x = -2$:

$f(-2) = 2(-2)^2 + 8(-2) - 1$

$f(-2) = 2(4) - 16 - 1$

$f(-2) = -9$

The vertex is the point (-2, -9).

The fourth feature of a quadratic function is the location of the $y$-intercept. Remember that the $y$-intercept of a function is the point where the function crosses the $y$-axis. The $x$-coordinate of this point is 0, and the $y$-intercept of the function is the point $(0, f(0))$. For a quadratic function $f(x) = ax^2 + bx + c$, the function evaluated at $x = 0$ is: $f(0) = a \times 0^2 + b \times 0 + c = c$.

In other words, the $y$-intercept of $f(x) = ax^2 + bx + c$ is the point $(0, c)$. The $y$-intercept is as easy to identify in a quadratic equation as it is when dealing with a linear function.

The fifth, and final, feature of a quadratic function has to do with the existence and the location of the $x$-intercepts of the function. Remember that the $x$-intercepts of a function are where the function intersects the $x$-axis. These points are found by setting the function equal to 0 and solving for $x$. In other words, finding the $x$-intercepts of the quadratic function $f(x) = ax^2 + bx + c$ involves solving the equation $ax^2 + bx + c = 0$. The three main methods for solving quadratic equations include factoring, completing the square, and using the quadratic formula.

Solving a quadratic equation of the form $ax^2 + bx + c = 0$, where $a$, $b$, and $c$ are integers, involves factoring the expression $ax^2 + bx + c$ and then setting each factor equal to 0. Factoring a quadratic expression of the form $ax^2 + bx + c$ involves looking at the factors of $a$ and $c$ and trying to find the right way to combine these factors to generate $b$. If $a$ is 1, the problem becomes one of finding two integers whose product is $c$ and whose sum is $b$. The signs of $a$ and $c$ indicate whether the integers will have the same sign or opposite sign. If the signs of $a$ and $c$ are the same, then the two integers will have the same sign. If the signs of $a$ and $c$ are the opposite of each other, then the two integers will have opposite signs. It then becomes trial and error to find the right integers. The more experience you have in factoring, the easier it becomes.

## Example 2

Solve the following quadratic equations by factoring:

a. $x^2 - 3x + 2 = 0$        c. $x^2 - 3x - 4 = 0$

b. $x^2 - 9 = 0$        d. $2x^2 - x - 3 = 0$

**Solution:**

a. Because the signs of $a$ and $c$ match, the two integers will have the same sign. We need to find two integers (with the same sign) whose sum is $-3$ and product is $+2$. Two integers that satisfy this condition are $-2$ and $-1$, and we have:
$x^2 - 3x + 2 = (x - 2)(x - 1)$.

Now take each factor, set it equal to 0, and solve:

$$(x - 2)(x - 1) = 0$$

| | |
|---|---|
| $(x - 2) = 0$ | $x - 1 = 0$ |
| $x = 2$ | $x = 1$ |

b. This is a special quadratic expression: the difference of two squares. These expressions are easy to factor:
$x^2 - 9 = (x - 3)(x + 3)$.
Now take each factor, set it equal to 0, and solve:

$$(x - 3)(x + 3) = 0$$

| | |
|---|---|
| $(x - 3) = 0$ | $(x + 3) = 0$ |
| $x = 3$ | $x = -3$ |

c. Because the signs of $a$ and $c$ are opposites, we need to find two integers (with opposite signs) whose product is $-4$ and whose sum is $-3$. Two integers that satisfy this condition are 1 and $-4$, and we have: $x^2 - 3x - 4 = (x - 4)(x + 1)$.
Now take each factor, set it equal to 0, and solve:

$$(x - 4)(x + 1) = 0$$

| | |
|---|---|
| $(x - 4) = 0$ | $(x + 1) = 0$ |
| $x = 4$ | $x = -1$ |

d. In this case, the leading coefficient is not 1, so we need to factor both $a$ and $c$ in order to factor the quadratic expression. After trial and error, we have: $2x^2 - x - 3 = (x + 1)(2x - 3)$.
Now take each factor, set it equal to 0, and solve:

$$(x + 1)(2x - 3) = 0$$

| | |
|---|---|
| $(x + 1) = 0$ | $(2x - 3) = 0$ |
| $x = -1$ | $x = \frac{3}{2}$ |

The second method for solving quadratic equations is called completing the square. Solving a quadratic equation of the form $ax^2 + bx + c = 0$ by completing the square involves focusing on the linear term involved in the expression. Suppose first that the leading coefficient is equal to 1. Rearrange the equation so that the variables are on one side and the constant term is on the other:

$$x^2 + bx + c = 0$$
$$x^2 + bx = -c$$

Next, add a number to both sides so that the expression on the left is a perfect square. Take the linear coefficient $b$, divide it in half, and then square the result: $\left(\frac{1}{2}b\right)^2 = \frac{1}{4}b^2$.

Add this number to both sides of the equation:

$$x^2 + bx + \frac{1}{4}b^2 = -c + \frac{1}{4}b^2$$

Now the expression on the left is a perfect square, which we can factor:

$$\left(x + \frac{1}{2}b\right)^2 = -c + \frac{1}{4}b^2$$

From here we can solve for $x$ by taking the square root of both sides of the equation and then subtracting $\frac{1}{2}b$:

$$\left(x + \frac{1}{2}b\right)^2 = -c + \frac{1}{4}b^2$$

$$\left(x + \frac{1}{2}b\right) = \pm\sqrt{\frac{1}{4}b^2 - c}$$

$$x = -\frac{1}{2}b \pm \sqrt{\frac{1}{4}b^2 - c}$$

## Example 3

Solve the quadratic equations by completing the square:

a. $x^2 + 4x - 2 = 0$          b. $x^2 - 3x - 4 = 0$

**Solution:**

a. Keep the terms involving variables on one side of the equation and move the constant over to the other side:

$$x^2 + 4x - 2 = 0$$

$$x^2 + 4x = 2$$

Take one-half of the coefficient in front of $x$ and square it:

$$\left(\frac{1}{2}4\right)^2 = 2^2 = 4. \text{ Add this to both sides of the equation:}$$

$$x^2 + 4x = 2$$

$$x^2 + 4x + 4 = 2 + 4$$

Now, the expression on the left is a perfect square:

$$x^2 + 4x + 4 = 6$$

$$(x + 2)^2 = 6$$

Finally, we can solve for $x$:

$(x + 2)^2 = 6$

$(x + 2) = \pm\sqrt{6}$

$x = -2 \pm\sqrt{6}$

b. Keep the terms involving variables on one side of the equation and move the constant over to the other side:

$x^2 - 3x - 4 = 0$

$x^2 - 3x = 4$

Take one-half of the coefficient in front of $x$ and square it:

$\left(\dfrac{1}{2}3\right)^2 = \dfrac{9}{4}$. Add this to both sides of the equation:

$x^2 - 3x = 4$

$x^2 - 3x + \dfrac{9}{4} = 4 + \dfrac{9}{4}$

Now, the expression on the left is a perfect square:

$x^2 - 3x + \dfrac{9}{4} = \dfrac{25}{4}$

$\left(x - \dfrac{3}{2}\right)^2 = \dfrac{25}{4}$

Finally, we can solve for $x$:

$\left(x - \dfrac{3}{2}\right)^2 = \dfrac{25}{4}$

$\left(x - \dfrac{3}{2}\right) = \pm\sqrt{\dfrac{25}{4}} = \pm\dfrac{5}{2}$

$x = \dfrac{3}{2} \pm \dfrac{5}{2}$

$x = \dfrac{3}{2} + \dfrac{5}{2} = \dfrac{8}{2} = 4$ or $x = \dfrac{3}{2} - \dfrac{5}{2} = \dfrac{-2}{2} = -1$

The third method for solving quadratic equations is to use the quadratic formula. The quadratic formula is a generalization of completing the square. The quadratic formula allows you to cut to the chase rather than working out all of the details involved in completing the square. The solutions to the equation $ax^2 + bx + c = 0$ are:

$$x = \frac{-b \pm \sqrt{b^2 - 4ac}}{2a}$$

## Example 4

Solve the following quadratic equations using the quadratic formula:

a. $x^2 + 4x - 2 = 0$  b. $x^2 - 6x - 5 = 0$

**Solution:**

a. In this case, $a = 1$, $b = 4$, and $c = -2$. Using the quadratic formula we have:

$$x = \frac{-b \pm \sqrt{b^2 - 4ac}}{2a}$$

$$x = \frac{-4 \pm \sqrt{4^2 - 4(1)(-2)}}{2(1)}$$

$$x = \frac{-4 \pm \sqrt{16 + 8}}{2}$$

$$x = \frac{-4 \pm \sqrt{24}}{2}$$

Now, we can factor 24: $24 = 4 \cdot 6$, and pull out the perfect square:

$$x = \frac{-4 \pm \sqrt{24}}{2}$$

$$x = \frac{-4 \pm 2\sqrt{6}}{2}$$

$$x = \frac{1}{2}\left(-4 \pm 2\sqrt{6}\right)$$

$$x = -2 \pm \sqrt{6}$$

b. In this case, $a = 1$, $b = -6$, and $c = -5$. Using the quadratic formula carefully we have:

$$x = \frac{-b \pm \sqrt{b^2 - 4ac}}{2a}$$

$$x = \frac{-(-6) \pm \sqrt{(-6)^2 - 4(1)(-5)}}{2(1)}$$

$$x = \frac{6 \pm \sqrt{36 + 20}}{2}$$

$$x = \frac{6 \pm \sqrt{56}}{2}$$

We can factor 56 as $56 = 4 \cdot 14$, and 4 is a perfect square:

$$x = \frac{6 \pm \sqrt{56}}{2}$$

$$x = \frac{6 \pm 2\sqrt{14}}{2}$$

$$x = \frac{1}{2}\left(6 \pm 2\sqrt{14}\right)$$

$$x = 3 \pm \sqrt{14}$$

Not all parabolas have $x$-intercepts. A parabola will have $x$-intercepts if the corresponding quadratic equation has a solution. You now have three methods available to solve quadratic equations. Realize that you can solve *any solvable* quadratic equation by using the quadratic formula. It is important to be able recognize whether or not a quadratic equation is solvable. One way to determine this is to evaluate the discriminant. Given the quadratic equation $ax^2 + bx + c = 0$, the **discriminant** is the value:

$$b^2 - 4ac$$

The quadratic equation will be solvable as long as the discriminant is *not* a negative number. Also, notice that if the discriminant is a perfect square then the solutions to the quadratic equation will be rational numbers (or maybe even whole numbers if we are lucky!). If the discriminant is a perfect square, then the square root of the discriminant will be a whole number. If $b$ and $a$ are also integers, then $x$ will be a ratio of integers (or a rational number).

Most of the time I will write a quadratic function as $f(x) = ax^2 + bx + c$. We can also write the equation of a parabola in standard form. The **standard form** of a parabola is an equation of the form $f(x) = a(x - h)^2 + k$. The advantage to writing a parabola in standard form is that the vertex can be identified immediately. Also, graphing a parabola in standard form involves the transformations discussed in Chapter 2. From the order of operations we can see the transformations involved and the order in which they occur. First there is a horizontal shift (because of the $(x - h)$ term). This is followed by a vertical stretch/contraction (because of the $a(x - h)$ term). There may be a reflection across the $x$-axis, if $a < 0$. Finally, there is a vertical translation (because of the $+k$ term). If you want to graph a parabola with equation $f(x) = a(x - h)^2 + k$, start with the parabola $f(x) = x^2$, shift it to the left/right $h$ units, stretch/contract it (depending on the magnitude of $a$), reflect it across the $x$-axis if $a < 0$, and then shift it up/down $k$ units.

The standard form of a parabola is very revealing: the axis of symmetry and the vertex can be read right off of the equation. Given a quadratic function written in the form $f(x) = ax^2 + bx + c$, we can write this quadratic function in standard form by completing the square. Start with the function $f(x) = ax^2 + bx + c$ and factor out the value of the leading coefficient, $a$, from the first two terms:

$$f(x) = a\left(x^2 + \tfrac{b}{a}x\right) + c.$$

Now add and subtract $\frac{b^2}{4a^2}$ inside of the parentheses (so that we are essentially adding 0 and not changing the function):

$$f(x) = a\left(x^2 + \tfrac{b}{a}x + \tfrac{b^2}{4a^2} - \tfrac{b^2}{4a^2}\right) + c.$$

The first three terms in parentheses can now be factored as a perfect square. The fourth term is in the way, so we will move it outside of the parentheses by multiplying it by the value $a$ that is just outside of the parentheses:

$$f(x) = a\left(x^2 + \tfrac{b}{a}x + \tfrac{b^2}{4a^2}\right) - \tfrac{b^2}{4a} + c.$$

Now factor the expression in parentheses:

$$f(x) = a\left(x + \tfrac{b}{2a}\right)^2 + \left(c - \tfrac{b^2}{4a}\right).$$

Compare this equation to the equation of a parabola in standard form:

$$f(x) = a(x - h)^2 + k.$$

Matching up the components, we see that

$$(x-h)=\left(x+\tfrac{b}{2a}\right),\text{ so } h=-\tfrac{b}{2a},\text{ and } k=\left(c-\tfrac{b^2}{4a}\right).$$

These formulas are hardly worth memorizing. If you remember the technique for completing the square, you can always go from the form $f(x) = ax^2 + bx + c$ to standard form.

## Example 5

Address the five features of the quadratic function $f(x) = x^2 - 2x - 5$ and write the function in standard form.

**Solution:**

a. The parabola is concave up, because the leading coefficient is positive.

b. The axis of symmetry is the line $x = \dfrac{-b}{2a} = \dfrac{-(-2)}{2(1)} = 1$.

c. The vertex is the point $(1, f(1))$, and
$f(1) = 1^2 - 2(1) - 5 = 1 - 2 - 5 = -6$,
so the vertex is the point $(1, -6)$.

d. The $y$-intercept is the point $(0, 5)$

e. The $x$-intercepts can be found by solving the equation
$x^2 - 2x - 5 = 0$. Using the quadratic formula, we have

$$x = \dfrac{-(-2)\pm\sqrt{2^2-4(1)(-5)}}{2(1)}$$

$$x = \dfrac{2\pm\sqrt{24}}{2}$$

$$x = \dfrac{2\pm2\sqrt{6}}{2}$$

$$x = 1\pm\sqrt{6}$$

The $x$-intercepts are the points $\left(1+\sqrt{6},0\right)$ and $\left(1-\sqrt{6},0\right)$.

f. The standard form of the quadratic function is found by completing the square:
$$f(x) = x^2 - 2x - 5$$

$$f(x) = (x^2 - 2x + 1 - 1) - 5$$
$$f(x) = (x^2 - 2x + 1) - 1 - 5$$
$$f(x) = (x - 1)^2 - 6$$

Quadratic functions can be used to model a variety of situations, including the trajectory of a ball thrown in the air and the revenue generated from the sale of a product.

### Example 6

A minor league baseball team plays in a park with a seating capacity of 10,000 spectators. With the ticket price set at $10, the average attendance at recent games has been 7,000. A market survey indicates that for each dollar the ticket price is lowered, the average attendance increases by 800 spectators. Using a linear function to model the attendance as a function of price, find a quadratic model for the *revenue* generated as a function of the price of a ticket. Use that model to find the ticket price that will maximize the revenue.

**Solution:** In Example 3 of the last lesson, we found a linear model for the attendance as a function of the price:

$$a = -800p + 15,000$$

We can use that model to create a function for the revenue generated from our ticket sales as a function of the price of the ticket. Let $R$ represent the revenue, and $p$ represent price. The revenue generated from ticket sales is found by taking the product of the number of tickets sold times the price per ticket. The number of tickets sold is the attendance, and the price per ticket is $p$. We can use our equation for attendance as a function of price to simplify this revenue equation:

$$R = a \cdot p$$
$$R = (-800p + 15,000) \cdot p$$
$$R = -800p^2 + 15,000p$$

Our equation for revenue is a parabola that is concave down, because the leading coefficient is negative. We can find the maximum possible revenue by finding the $p$-coordinate of the vertex of the parabola:

$$p = \frac{-(15{,}000)}{2(-800)} = 9.375$$

We can round our answer to $9.38.

## Lesson 3-2 Review

Address the five features of the following quadratic functions, and write the functions in standard form.

1. $f(x) = x^2 - 8x + 12$                 2. $f(x) = -x^2 + 6x - 4$

# Lesson 3-3: Quadratic Inequalities

A quadratic equation is an equation of the form $ax^2 + bx + c = 0$. A quadratic inequality occurs when a quadratic function is less than 0, less than or equal to 0, greater than 0, or greater than or equal to 0. Solving a quadratic inequality makes use of a very important property of quadratic functions called continuity.

A quadratic function can have either two, one, or no real zeros. The zeros of a quadratic function are the only place where a quadratic function can change sign. A change in sign means that the function is positive and then becomes negative, or vice versa. In other words, if a quadratic function is positive at one point, then it will stay positive until it reaches one of its zeros. If a quadratic function has no zeros, the function has no $x$-intercepts, and the function cannot ever change sign. To simplify, if a quadratic function is positive at one point, it will be positive on its entire domain. The function $f(x) = x^2 + 1$ has no real zeros, and this function is positive on its entire domain.

If a quadratic function has one zero, it also will never change sign. The zero of the quadratic function represents the point where the quadratic function just touches the $x$-axis. After that, it turns around and heads in the opposite direction. The function $f(x) = x^2$ has one real zero (at $x = 0$), and this function is never negative. The function decreases until it just touches the $x$-axis at $x = 0$, and then it turns around and heads in the opposite direction.

If a quadratic function has two zeros, it will cross the $x$-axis at both of its zeros. The quadratic function will change sign at both of the zeros. The function $f(x) = x^2 - 1$ has two real zeros (at $x = 1$ and $x = -1$), and the function crosses the x-axis at these points. The function is above the x-axis on the interval $(-\infty, -1)$, and then it is below the axis on the interval $(-1, 1)$, and then it is above the x-axis on the interval $(1, \infty)$. The graph of $f(x) = x^2 - 1$ is shown in Figure 3.3.

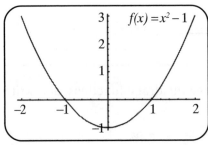

Figure 3.3

All quadratic functions have the property that they can only change signs at their x-intercepts. We will make use of this property of quadratic functions when we solve quadratic inequalities. We will solve all quadratic inequalities using the same method. I will walk you through the solution to the quadratic inequality $x^2 - 2x - 8 > 0$ as I present the general procedure.

1. Make sure that 0 is on one side of the inequality. For the inequality $x^2 - 2x - 8 > 0$, this condition is met.

2. Find the zeros of the quadratic *function* that appears in the inequality:

$$x^2 - 2x - 8 = 0$$

$$(x - 4)(x + 2) = 0$$

$$x = 4 \text{ or } x = -2$$

3. Arrange the zeros of the quadratic function on a number line, as shown in Figure 3.4.

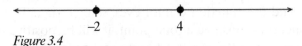

Figure 3.4

4. The two zeros of the quadratic function break the number line into three pieces. Select one test value from each of the three regions, and add these test values to the number line. Circle each test value, to distinguish them from the zeros of the quadratic function. For our problem, I will choose the test values −3, 0, and 5. Keep in mind that you can choose *any* number in each of the three regions.

5. Evaluate the *sign* of the quadratic function at each of these test values. There is no need to actually evaluate the function, just the sign of the

| x | Sign of $(x-4)$ | Sign of $(x+2)$ | Sign of $(x-4)(x+2)$ |
|---|---|---|---|
| −3 | − | − | + |
| 0 | − | + | − |
| 5 | + | + | + |

function. The easiest way to evaluate the sign of the function is to use the factored form of the function. I've outlined the logic of determining the sign of the function in the table on page 71.

6.  Use the table to determine the sign of the quadratic function in each of the three regions. The signs of the quadratic function are shown in Figure 3.5.

*Figure 3.5*

7.  Answer the question. In this problem, we were asked to find the regions where $x^2 - 2x - 8 > 0$. From Figure 3.5, we can see that $x^2 - 2x - 8 > 0$ on the regions $(-\infty, -2) \cup (4, \infty)$.

The solution to a quadratic inequality will include the zeros of the function if the inequality is $\geq$ or $\leq$. In this case, we would use brackets (either [ or ]) next to the zeros of the quadratic function. If the inequality is strict (meaning > or <) we use parentheses next to the zeros of the quadratic function.

## Example 1

Solve the quadratic inequality $x^2 - x \leq 6$.

**Solution:** Follow the procedure outlined.

1.  Make sure that 0 appears on one side of the inequality. Subtract 6 from both sides of the inequality: $x^2 - x - 6 \leq 0$.

2.  Find the zeros of the quadratic function: 
$$x^2 - x - 6 = 0$$
$$(x - 3)(x + 2) = 0$$
$$x = 3 \text{ or } x = -2$$

3.  The number line is shown in Figure 3.6.

*Figure 3.6*

4.  Use the test values $x = -3$, $x = 0$, and $x = 4$.

5.  The sign chart is shown in the adjacent table.

| $x$ | Sign of $(x-3)$ | Sign of $(x+2)$ | Sign of $(x-3)(x+2)$ |
|---|---|---|---|
| $-3$ | $-$ | $-$ | $+$ |
| $0$ | $-$ | $+$ | $-$ |
| $4$ | $+$ | $+$ | $+$ |

6. The signs of the quadratic function are shown in Figure 3.7.

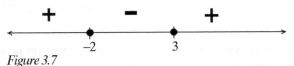

*Figure 3.7*

7. From Figure 3.7, we can see that $x^2 - x \le 6$ in the region $[-2, 3]$.

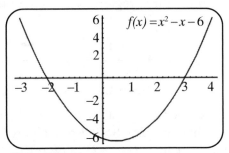

*Figure 3.8*

The graph of $f(x) = x^2 - x - 6$ is shown in Figure 3.8. The regions where the graph of this function touches or is below the $x$-axis indicate where $x^2 - x - 6 \le 0$.

---

Another method that can be used to determine the solutions to a quadratic inequality involves graphing each of the functions and determining the regions where one function is lower than the other. A graphing calculator is particularly useful in checking your algebraic solution. It is important to be able to solve quadratic inequalities algebraically as well as graphically. Keep in mind that a graphing calculator only shows you the graph of a function in a specific window. The function could have some interesting twists and turns that may occur outside of the default window of the graphing calculator, but you won't miss any of the important information if you solve these problems algebraically. Use a graphing calculator to check your work, but be sure to understand the algebra behind the method. The graph of the functions $f(x) = x^2 - x$ and $g(x) = 6$ are shown in Figure 3.9. The inequality $x^2 - x \le 6$ is satisfied when the parabola $f(x) = x^2 - x$ lies below the horizontal line $g(x) = 6$.

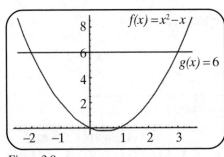

*Figure 3.9*

### Lesson 3-3 Review

Solve the following quadratic inequalities:

1. $8x - 12 \ge x^2$

2. $x^2 - 3x < 4$

# Lesson 3-4: Piecewise-Defined Functions

A piecewise-defined function is a function that is defined by different formulas on different parts of its domain. When evaluating a piecewise-defined function, you must determine which formula to use based on the region of the domain that contains the particular value of the independent variable.

One piecewise-defined function that appears frequently in mathematics is the absolute value function. The absolute value of a number can never be negative. The absolute value of a positive number is itself, and the absolute value of a negative number is the resulting positive number obtained by multiplying the negative number by −1. The absolute value of a number can

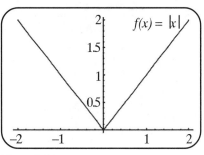

*Figure 3.10*

be written as: $|x| = \begin{cases} -x & x < 0 \\ x & x \geq 0 \end{cases}$ and its graph is shown in Figure 3.10.

To graph a piecewise-defined function, first break the domain into the pieces indicated by the inequalities. Graph each function in its appropriate piece; be careful not to extend any of the graphs beyond the boundary of the regions. The graph of the piecewise-defined

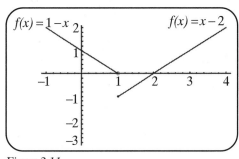

*Figure 3.11*

function $f(x) = \begin{cases} 1-x & x < 1 \\ x-2 & x \geq 1 \end{cases}$ is shown in Figure 3.11.

There are some observations to make. First of all, notice that at the boundary $x = 1$, the two pieces of the function do not meet. One of the pieces ends with an open circle, and the other ends with a closed circle. The side of the domain with the strict inequality has the open circle; if one of the sides is either ≥ or ≤, then its graph will end at the boundary and have a closed circle. The closed circle indicates that if a point lies on the boundary of two regions, the value of the function is determined by the rule that does *not* involve a strict inequality. If both sides involve a strict inequality, then both sides will have open circles at the boundary.

In general, the domain of a function can be broken up into many pieces, but the pieces *cannot* overlap. We cannot have two different rules apply to a particular input value. Piecewise-defined functions are often used to determine cell phone bills, utility bills, income tax, royalties on book sales, or situations in which a discount is given for a bulk purchase above a threshold amount.

## Lesson 3-4 Review

Graph the following piecewise-defined functions:

1. $f(x) = \begin{cases} 2-x & x<0 \\ 2x & x\geq0 \end{cases}$

2. $f(x) = \begin{cases} -1 & x<0 \\ 1 & x\geq0 \end{cases}$

## Answer Key

### Lesson 3-2 Review

1. $f(x) = x^2 - 8x + 12$:

    it opens up,
    the axis of symmetry is $x = 4$,
    the vertex is $(4, -4)$,
    the $y$-intercept is $(0, 12)$,
    the $x$-intercepts are $(6, 0)$ and $(2, 0)$,
    and the standard form of this parabola is $y = (x-4)^2 - 4$.

2. $f(x) = -x^2 + 6x - 4$:

    it opens down,
    the axis of symmetry is $x = 3$,
    the vertex is $(3, 5)$,
    the $y$-intercept is $(0, -4)$,
    the $x$-intercepts are $(3 + \sqrt{5}, 0)$ and $(3 - \sqrt{5}, 0)$,
    and the standard form of this parabola is $y = -(x-3)^2 + 5$.

### Lesson 3-3 Review

1. $8x - 12 \geq x^2$: $[2, 6]$
2. $x^2 - 3x < 4$: $(-1, 4)$

## Lesson 3-4 Review

1.

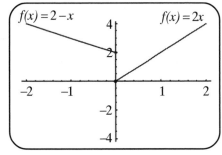

*Figure 3.12*

The graph of $f(x) = \begin{cases} 2-x & x < 0 \\ 2x & x \geq 0 \end{cases}$

2.

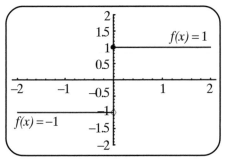

*Figure 3.13*

The graph of $f(x) = \begin{cases} -1 & x < 0 \\ 1 & x \geq 0 \end{cases}$

# Polynomials

Linear and quadratic functions are examples of a general class of functions called polynomials. Polynomials can be used to model real-world situations. Polynomials have been used in encryption and appear in solutions to problems in quantum mechanics.

## Lesson 4-1: Characteristics of Polynomials

A **polynomial** is a function of the form:

$$f(x) = a_n x^n + a_{n-1} x^{n-1} + \ldots + a_1 x + a_0.$$

The coefficients $a_n, a_{n-1}, \ldots, a_1, a_0$ are real numbers with $a_n \neq 0$. The constant $a_n$ is called the **leading coefficient**, and $a_0$ is referred to as the **constant coefficient**, or the **constant term**. The number $n$ must be a non-negative integer, and is called the **degree** of the polynomial. The domain of a polynomial function is the set of all real numbers.

The graph of a polynomial is related to its degree. We will expand on the method for solving quadratic inequalities to develop a graphing strategy for graphing polynomials. Before we learn to graph some specific types of polynomials, there are a few quick observations to make about the graphs of polynomials in general.

Figure 4.1 on page 78 shows the graphs of some polynomials of increasing degree. The first polynomial has degree 1, the second has degree 2, and so on. The degree of the sixth polynomial is 6. Notice that the polynomials with odd degree, share some common characteristics, and the polynomials with even degree also have common properties. We will examine these graphs in more detail.

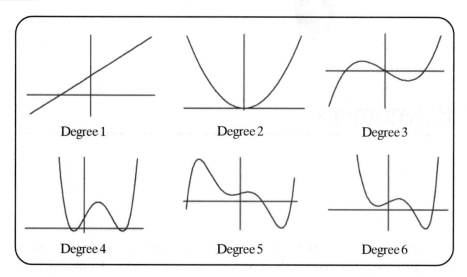

*Figure 4.1*

None of these polynomials have no breaks in their graphs. Polynomials are examples of functions that are **continuous**. Polynomials are also **smooth**, in that there are no sharp corners or cusps. The polynomials of odd degree all cross the $x$-axis. If they start out negative, they eventually end up positive, or vice versa. A polynomial of odd degree may or may not have a **turning point**, or a point where the function changes direction from increasing to decreasing, or vice versa. The polynomials of even degree may or may not cross the $x$-axis. If they start out negative, they eventually end up negative, and if they start out positive, they eventually end up positive. Notice that each of the polynomials of even degree has at least one turning point. In general, a polynomial of degree $n$ will cross the $x$-axis at most $n$ times, and will have at most $(n - 1)$ turning points.

Polynomials can also exhibit certain symmetries. If a polynomial only involves odd powers of $x$ and passes through the origin, so that $a_0 = 0$, then the graph of the polynomial will be symmetric about the origin. In other words, a polynomial that only involves *odd* powers and has a constant coefficient equal to 0 is an *odd* function. Similarly, if a polynomial only involves even powers of $x$, then the graph of the polynomial will be symmetric about the $y$-axis. To simplify, a polynomial that only involves *even* powers is an *even* function. An even function does not have a restriction on the constant coefficient; $a_0$ does not have to be 0 for an even function, because even functions do not have to pass through the origin. Combining even and odd powers of $x$ destroys the even/odd symmetry of a polynomial. A polynomial that involves both even and odd powers of $x$ will not be even, because of the odd powers. It will not be odd, because of the even powers. There is only *one* polynomial that is both even and odd: the function $f(x) = 0$.

## Lesson 4-1 Review

1. Determine whether the following polynomials are even, odd, or neither.

a. $f(x) = 3x^4 - 2x + 1$  c. $f(x) = 3x^9 - 2x^5 + 1$

b. $f(x) = 3x^{10} - 2x^6 + 5$  d. $f(x) = 3x^9 - 2x^5 + x$

# Lesson 4-2: Dividing Polynomials

The process of dividing one polynomial by another polynomial is similar to long division with natural numbers. Because you are working with polynomials rather than natural numbers, you must make use of the degree of the polynomials involved in the division. A systematic approach to division is as follows:

1. Arrange the terms of both polynomials in descending order according to their degrees. If the polynomial has gaps between consecutive terms, leave a space.

2. Determine the monomial that you need to multiply the first term of the divisor by in order to get the first term in the dividend. Place this monomial above the long division symbol.

3. Multiply the entire divisor by this monomial and place this answer directly below the dividend. Line up like terms.

4. Subtract this new polynomial from the dividend.

5. If the degree of the result from the subtraction is less than the degree of the divisor, you are finished. If it is not, repeat steps 2-4 until you are left with a polynomial whose degree is smaller than the degree of the divisor. The polynomial above the long division symbol is called the *quotient*, and the leftover polynomial is called the *remainder*.

If the remainder is zero, then the division is exact, and the quotient is the polynomial written across the top of the long division symbol. If the remainder is not zero, then the quotient is the polynomial written across the top of the long division symbol plus the remainder divided by the divisor.

## Example 1

Find the following:

a. $\dfrac{6x^2+12x+1}{2x+3}$

b. $\dfrac{x^3+1}{x+1}$

**Solution:**

a.
$$2x+3\overline{\smash{\big)}6x^2+12x+1} \quad\quad 3x+\tfrac{3}{2}$$

$$\underline{(-)6x^2+9x}$$
$$3x+1$$
$$\underline{(-)\,3x+\tfrac{9}{2}}$$
$$-\tfrac{7}{2}$$

The degree of the remainder is 0 (the remainder is not zero, but it's degree is!), which is less than the degree of the divisor, so you are done. We see that

$$\frac{6x^2+12x+1}{2x+3}=\left(3x+\frac{3}{2}\right)-\frac{\tfrac{7}{2}}{2x+3}.$$

b.
$$x+1\overline{\smash{\big)}x^3+0x^2+0x+1} \quad\quad x^2-x+1$$

$$\underline{(-)x^3+x^2}$$
$$-x^2+1$$
$$\underline{(-)-x^2-x}$$
$$x+1$$
$$\underline{(-)\,x+1}$$
$$0$$

Notice that the remainder is 0, and $\dfrac{x^3+1}{x+1}=x^2-x+1.$

If the remainder is 0, then the polynomial in the denominator is a factor of the polynomial in the numerator. From the result in Example 1b, we can multiply both sides of the equation $\frac{x^3+1}{x+1} = x^2 - x + 1$ by $(x + 1)$, and see that $x^3 + 1 = (x + 1)(x^2 - x + 1)$. This is the key to factoring a polynomial.

Suppose we want to find the quotient $\frac{P(x)}{(x-c)}$. The degree of the divisor is 1, so the degree of the remainder will be 0. In other words, the remainder will be a constant. We can write

$$\frac{P(x)}{(x-c)} = Q(x) + \frac{R}{(x-c)},$$

where $R$ is a constant. If we multiply both sides of this equation by $(x - c)$, then we have the equation:

$$P(x) = (x - c)Q(x) + R.$$

We can find the remainder, $R$, by evaluating $P(x)$ at $x = c$:

$$P(c) = (c - c)Q(c) + R.$$

It doesn't matter what the value of $Q(c)$ is, because we are multiplying it by $(c - c)$, which is 0. In other words, the remainder, $R$, can be found by the equation $R = P(c)$.

This provides us with a shortcut method to find the remainder of the quotient $\frac{P(x)}{(x-c)}$. Just evaluate the polynomial in the numerator at $x = c$!

## Example 2

Find the remainder of the quotient $\dfrac{x^3 + 2x + 1}{x - 3}$.

**Solution:** Although we certainly could do the division, it is much easier to evaluate the polynomial in the numerator at $x = 3$. The remainder of $\dfrac{x^3 + 2x + 1}{x - 3}$ is $3^3 + 2(3) + 1 = 34$. If you do the long division, you will find that the remainder is, in fact, 34.

This method only works when the degree of the divisor is 1. The reason for this is that the degree of the remainder will always be less than the degree of the divisor. If the degree of the divisor is 1, the degree of the remainder must be 0, so the remainder must be a constant. That is important. To find the remainder of $\frac{x^{50}+2x^2+1}{x^2+1}$, you must do the division!

The remainder of the quotient $\frac{x^{200}-3x^{48}-2x+4}{x+1}$ is just the numerator evaluated at $x = -1$. In general, the remainder of the quotient $\frac{P(x)}{(x-c)}$ is $P(c)$. This result is known as the Remainder Theorem.

> **Remainder Theorem: If $R$ is the remainder when a polynomial $P(x)$ is divided by $(x - c)$, then $R = P(c)$.**

There is a shortcut to long division, but this shortcut only works if the divisor is of the form $(x - c)$. This method is called synthetic division, and one advantage of this method is that it saves a lot of writing. There is one catch: you must include *all* of the coefficients of the polynomial in the numerator, including the ones that are invisible because they are 0. If you forget that important condition, your work will be flawed. The best way to illustrate this method is to work out an example.

## Example 3

Find the quotient $\frac{3x^3-5x-12}{x-2}$.

**Solution:** First of all, write the opposite of the constant term in the divisor to the left, and put half of a box around it, to keep it separate from the other constants we will write. In this case, the constant in the denominator is $-2$, so our constant in the half-box will be 2. Next, write out all of the coefficients of the polynomial, in order.

2⌋ 3  0  −5  −12

Next, take the leading coefficient of the polynomial (3) and bring it down below the horizontal line.

2⌋ 3  0  −5  −12

————————

3

Multiply the number below the horizontal line (3) by the number in the half-box (2) and write the result (6) below the next coefficient (0).

$$2\rfloor \quad 3 \quad 0 \quad -5 \quad -12$$
$$6$$

$$3$$

Add the numbers in the second column together, and write the sum directly below the numbers you just added.

$$2\rfloor \quad 3 \quad 0 \quad -5 \quad -12$$
$$6$$

$$3 \quad 6$$

Repeat this process with the next set of numbers. Multiply the number below the horizontal line (6) by the number in the half-box (2) and write the result above the horizontal line directly underneath the next coefficient.

$$2\rfloor \quad 3 \quad 0 \quad -5 \quad -12$$
$$6 \quad 12$$

$$3 \quad 6$$

Add these two numbers together, and continue the process until you have worked with each of the coefficients of the polynomial:

$$2\rfloor \quad 3 \quad 0 \quad -5 \quad -12$$
$$6 \quad 12 \quad 14$$

$$3 \quad 6 \quad 7 \quad 2$$

The last step is to interpret your answer. The last number below the horizontal line is the remainder after doing the division. The other three numbers are the coefficients of the quotient. In other words,

$$\frac{3x^3 - 5x - 12}{x - 2} = 3x^2 + 6x + 7 + \frac{2}{x - 2}.$$

When evaluating quotients using synthetic division, the degree of the quotient will always be one less than the degree of the polynomial in the numerator. We can easily check the remainder obtained using synthetic division with the remainder that we expect from the Remainder Theorem. For the quotient $\frac{3x^3 - 5x - 12}{x - 2}$, the remainder should be the value of the polynomial in the numerator evaluated at $x = 2$:

$3(2)^3 - 5(2) - 12 = 24 - 10 - 12 = 2.$

This agrees with the last number underneath the horizontal line in the synthetic division example we just worked.

I mentioned earlier that if, after dividing a polynomial $P(x)$ by the linear factor $(x - c)$, the remainder is 0, then we say that $(x - c)$ is a factor of $P(x)$. Let $Q(x)$ represent the quotient $\frac{P(x)}{(x-c)}$. In other words, $\frac{P(x)}{(x-c)} = Q(x)$. This equation is equivalent to $P(x) = (x - c)Q(x)$. This idea is crucial in understanding how to factor polynomials, which we will discuss in detail in the next lesson.

### Lesson 4-2 Review

Find the remainder of the following:

1. $\dfrac{x^{30} - 2x^5 + 3x - 1}{x + 1}$

2. $\dfrac{x^4 - 2x^2 - 8}{x^2 + 2}$

3. If a polynomial $P(x)$ passes through the points $(2, 3)$ and $(-2, 5)$, find the remainder of $\dfrac{P(x)}{x + 2}$.

## Lesson 4-3: Zeros of Polynomials

Understanding the behavior of a polynomial involves thinking about polynomials in a more abstract sense. In this chapter, we will only work with polynomials with integer coefficients. Every polynomial of degree $n$ has exactly $n$ zeros. These zeros do not have to be distinct. For example, the polynomial $P(x) = x^2$ has two zeros, but the two zeros are not distinct.

The $x$-intercepts of a polynomial are also called the zeros of a polynomial or the roots of a polynomial. The zeros of a polynomial are the values of $x$ that, when substituted into the formula for the polynomial, give a value of 0. The zeros of a quadratic function can be found by using the quadratic formula. The zeros of a polynomial of degree 3 (a cubic polynomial), or higher, are not as easy to find. However, if a polynomial is factored completely, it is easy to determine the zeros of the polynomial: set each factor equal to 0 and solve for $x$.

## Example 1

Find the zeros of the following polynomials:

a. $P(x) = (x - 3)(x + 2)(2x + 1)$

b. $Q(x) = x(x - 4)^2(3x - 1)$

**Solution:** Set each factor equal to 0 and solve for x:

a. $P(x) = 0$ when $x = 3, x = -2$ and $x = -\dfrac{1}{2}$.

b. $Q(x) = 0$ when $x = 0, x = 4$ and $x = \dfrac{1}{3}$.

_____

The zeros of a polynomial are important points. They are the only places where the polynomial can change sign. Once you know all of the zeros of a polynomial, you know almost everything about the polynomial. The factors of a polynomial and the zeros of a polynomial are related to each other through the **Factor Theorem.**

**The Factor Theorem: $c$ is a zero of a polynomial $P(x)$ if, and only if, $(x - c)$ is a factor of $P(x)$.**

This is an important theorem because it allows us to find the zeros of a polynomial directly from the factors of the polynomial, and vice versa. For example, if $P(x)$ is a polynomial with $P(3) = 0$, then $(x - 3)$ is a factor of $P(x)$, and we can write $P(x) = (x - 3)Q(x)$, where $Q(x)$ is a polynomial whose degree is one less than $P(x)$. It is this process that enables us to factor polynomials completely by searching for its zeros, or looking for where the polynomial crosses the $x$-axis. Once we find a zero of $P(x)$, we can break $P(x)$ into the product of a linear factor, $(x - c)$, and a polynomial of smaller degree, $Q(x)$. Remember that if the remainder of $\dfrac{P(x)}{(x-c)}$ is 0, we can write $\frac{P(x)}{(x-c)} = Q(x)$, where $Q(x)$ is a polynomial whose degree is one less than the degree of $P(x)$. We can rearrange this equation to obtain the equation $P(x) = (x - c)Q(x)$. Now we can turn our attention to $Q(x)$: find one place where the graph of $Q(x)$ crosses the $x$-axis. Once we find a zero of $Q(x)$, we can break $Q(x)$ into the product of a linear factor and a polynomial of even smaller degree. Continuing in this manner, we have a method for factoring polynomials.

More often than not, the zeros of a polynomial are irrational numbers that are difficult to determine precisely. Fortunately, the rational zeros of a polynomial are easy to spot. At first glance, it may seem that *any* rational number can be a zero of a polynomial. Actually, given any polynomial with integer coefficients, there are only a handful of rational numbers that can possibly be zeros of the polynomial. The rational zeros of a polynomial are closely related to the leading coefficient and the constant term in a polynomial. This relationship is described in the **Rational Zero Theorem**.

**Rational Zero Theorem: If $P(x) = a_n x^n + a_{n-1} x^{n-1} + ... + a_1 x + a_0$ is a polynomial with integer coefficients, with $a_n \neq 0$ and $a_0 \neq 0$, and $\frac{p}{q}$ (written in lowest terms) is a rational zero of $P(x)$, then p is a factor of $a_0$ and q is a factor of $a_n$.**

This theorem is very useful in finding the rational zeros of a polynomial. For example, we can see right away that $x = 3$ cannot possibly be a rational zero of the polynomial $P(x) = 2x^4 - x^3 - 42x^2 - 19x + 20$, because 3 is not a factor of the constant term of the polynomial (20). In fact, we can write out all of the rational numbers that could possibly be zeros of this polynomial. First, find all possible factors (positive and negative) of the constant term. Remember that every integer is evenly divisible by 1 and itself. So this list shouldn't be empty, even if the number is prime! Next, find all possible factors (positive and negative) of the leading coefficient. Find all possible ratios of the factors of the constant term divided by factors of the leading term. That list of numbers contains the only rational numbers that could possibly be zeros of the polynomial. If none of those numbers in that list are zeros of the polynomial, then the zeros of the polynomial will either be irrational numbers, or they will be complex numbers.

## Example 2

Write out the list of possible rational zeros of the polynomial
$P(x) = 2x^4 - x^3 - 42x^2 - 19x + 20$.

Is $x = \dfrac{7}{2}$ a possible zero of this polynomial?

**Solution:** Write out the possible factors of 20: ±1, ±2, ±4, ±5, ±10, ±20. Next, write out the possible factors of 2: ±1, ±2. Finally, take all possible combinations of the numbers in the first list divided by the numbers in the second list. Start out systematically. Take the numbers in the first list and divide them all by the first number in the second list (1) first. Then take the numbers in the first list and divide them by the second number in the list (2), and throw out any duplicates:

The list of possible zeros are: ±1, ±2, ±4, ±5, ±10, ±20,

$\pm\dfrac{1}{2}, \pm\dfrac{2}{2}, \pm\dfrac{4}{2}, \pm\dfrac{5}{2}, \pm\dfrac{10}{2}, \pm\dfrac{20}{2}$. There are some duplicates on this list:

$\pm\dfrac{2}{2} = \pm1$, $\pm\dfrac{4}{2} = \pm2$, $\pm\dfrac{10}{2} = \pm5$, and $\pm\dfrac{20}{2} = \pm10$. The duplicates can

be thrown out. Our compact list of possible rational zeros is ±1,

±2, ±4, ±5, ±10, ±20, $\pm\dfrac{1}{2}, \pm\dfrac{5}{2}$. Notice that $\dfrac{7}{2}$ is not on this list.

That's because 7 is not a factor of 20.

Therefore, $x = \dfrac{7}{2}$ is not a possible zero of this polynomial.

---

The zeros of a polynomial with integer coefficients must fall into one of three categories. A zero can be a rational number, an irrational number, or a complex number whose imaginary component is not zero. These three categories are mutually exclusive. We already have a strong understanding of the rational zeros, and we can actually list all of the possible rational zeros of any polynomial with integer coefficients. The irrational zeros are more difficult to find. And the complex zeros have some interesting characteristics, which I will now discuss.

Before I talk about the complex zeros of a polynomial, I need to briefly discuss complex numbers. As you probably know, there is no real number whose square is −1. Mathematicians needed a number whose square is −1, so they invented it, and called it $i$: $i$ is the number whose square is −1: $i^2 = -1$. A complex number has two components: a real component and an imaginary component. Any complex number can be written in the form $a + bi$, where $a$ is called the real part of the complex number and $b$ is called the imaginary part of the complex number. Keep in mind that the imaginary part of a complex number is actually a real number!

Every complex number has a *complex conjugate*. The complex conjugate of the number $a + bi$ is $a - bi$. A complex number and its conjugate have a special relationship. When a complex number is multiplied by its complex conjugate, the result is a positive real number:

$$(a + bi)(a - bi) = a^2 + b^2.$$

If the complex number $a + bi$ is a zero of a polynomial with integer coefficients, then its complex conjugate is also a zero of that polynomial. In other words, complex zeros travel in (conjugate) pairs, so if a polynomial with real coefficients has complex zeros, it must have an even number of them. Remember that a degree n polynomial must have exactly n zeros. These zeros can be rational, irrational, or complex. The restriction for the rational zeros is that they must be on our list of possible rational zeros. The restriction for the complex zeros is that they must travel in pairs, or there must be an even number of them.

## Example 3

Is it possible for a degree 3 polynomial with integer coefficients to have zeros at $x = 2 + i$ and $x = 3 - 4i$?

**Solution:** No, it is not. A degree 3 polynomial with integer coefficients will have exactly three zeros. Because complex zeros travel in conjugate pairs, if the polynomial had zeros at $x = 2 + i$ and $x = 3 - 4i$, it would also have to have zeros at $x = 2 - i$ and $x = 3 + 4i$. That means that the degree 3 polynomial would have four zeros, which is not possible.

---

Another consequence of the fact that complex zeros travel in pairs is that every polynomial whose degree is odd must have at least one real zero. A polynomial with odd degree will have an odd number of zeros. There can only be an even number of complex zeros, which leaves at least one zero unaccounted for. That one unaccounted-for zero must be real.

## Lesson 4-3 Review

1. Write the list of possible rational zeros of the polynomial $6x^5 + 3x^4 - 2x^2 + 3x - 8$.

2. Can $(x - 7)$ be a factor of $6x^4 + 29x^3 - 24x^2 - 101x - 30$? Explain your answer.

# Lesson 4-4: Factoring Polynomials

The strategy for factoring a degree n polynomial, $P(x)$, is to start with the list of possible rational zeros. Every list will start off with ±1. Use synthetic division to quickly see whether the rational numbers in the list work. In other words, divide $P(x)$ by $(x - c)$ where $c$ is a rational number in the list. If $(x - c)$ is a factor of $P(x)$, write the polynomial as $P(x) = (x - c)Q(x)$, where $Q(x)$ is a polynomial with degree n–1. Now apply our factoring techniques to $Q(x)$. Come up with a new list of possible rational zeros, and start trying them. With this systematic approach, you will be able to factor any polynomial with rational zeros. Every time you find a rational zero, the degree of the polynomial that you are trying to factor decreases by 1. Eventually, you will end up with a quadratic function, and you can find the zeros of a quadratic function using the quadratic formula. Remember that once you find a number $c$ that works, the factor is written as $(x - c)$.

## Example 1

Factor the polynomial $x^3 - x^2 - 5x - 3$.

**Solution:** Find the list of possible rational zeros: the factors of 3 are ±1 and ±3. The factors of 1 are ±1. From this, we see that the only possible rational zeros are ±1 and ±3. Use synthetic division to try each of the factors. I recommend starting with ±1.

$$
\begin{array}{r|rrrr}
1 & 1 & -1 & -5 & -3 \\
  &   & 1 & 0 & -5 \\
\hline
  & 1 & 0 & -5 & -8
\end{array}
\qquad
\begin{array}{r|rrrr}
-1 & 1 & -1 & -5 & -3 \\
   &   & -1 & 2 & 3 \\
\hline
   & 1 & -2 & -3 & 0
\end{array}
$$

Because –1 worked, we can factor $x^3 - x^2 - 5x - 3$ into $(x - (-1))$ $(x^2 - 2x - 3)$. At this point, we have a quadratic function that we can factor easily: $x^3 - x^2 - 5x - 3 = (x + 1)(x - 3)(x + 1)$.

## Example 2

Factor the polynomial $P(x) = 2x^4 - x^3 - 42x^2 - 19x + 20$.

**Solution:** Find the list of possible rational zeros. We found this list in Example 2 of Lesson 4-3: $\pm 1, \pm 2, \pm 4, \pm 5, \pm 10, \pm 20, \pm\dfrac{1}{2}, \pm\dfrac{5}{2}$.

Use synthetic division to try each of the factors, starting with $\pm 1$.

| $1 \rfloor$ | 2 | -1 | -42 | -19 | 20 |
|---|---|---|---|---|---|
| | | 2 | 1 | -41 | -60 |
| | 2 | 1 | -41 | -60 | -40 |

| $-1 \rfloor$ | 2 | -1 | -42 | -19 | 20 |
|---|---|---|---|---|---|
| | | -2 | 3 | 39 | -20 |
| | 2 | -3 | -39 | 20 | 0 |

Because −1 worked, we can factor $P(x) = 2x^4 - x^3 - 42x^2 - 19x + 20$ into $(x -(-1))(2x^3 - 3x^2 - 39x + 20) = (x + 1)(2x^3 - 3x^2 - 39x + 20)$. Now we can focus on factoring the new polynomial, $2x^3 - 3x^2 - 39x + 20$. Because this polynomial has the same constant coefficient and the same leading coefficient as our original polynomial, our list of possible rational zeros will not change. Because $(x - 1)$ was not a factor of the original polynomial, it cannot be a factor of our new polynomial. Because $(x + 1)$ was a factor of the original polynomial, it could still be a factor of this new polynomial. If a factor worked once, it could work twice, or even three times! But once a factor has been ruled out, it remains ruled out throughout the problem. We need to try −1 again.

| $-1 \rfloor$ | 2 | -3 | -39 | 20 |
|---|---|---|---|---|
| | | -2 | 5 | 34 |
| | 2 | -5 | -34 | 54 |

The remainder is 54, not 0, so −1 doesn't work again. Now we have ruled out 1 and −1 for the rest of this problem. Moving down the list, we can try $\pm 2$ (neither of which work), and $\pm 4$. As it turns out, −4 works:

| $-4 \rfloor$ | 2 | -3 | -39 | 20 |
|---|---|---|---|---|
| | | -8 | 44 | -20 |
| | 2 | -11 | 5 | 0 |

Our polynomial is now factored as $(x + 1)(x + 4)(2x^2 - 11x + 5)$. The polynomial that we have to factor next is $2x^2 - 11x + 5$, which is a quadratic function. We should be able to factor it using other methods. If we wanted to stick with this method, notice that the constant term has changed from the original polynomial: it started out as 20, and now it is 5. So our original list can be shortened drastically. Instead of having $\pm 1$, $\pm 2$, $\pm 4$, $\pm 5$, $\pm 10$, $\pm 20$, $\pm \dfrac{1}{2}$, $\pm \dfrac{5}{2}$, we would only have $\pm 1$, $\pm 5$, $\pm \dfrac{1}{2}$, $\pm \dfrac{5}{2}$. The other terms, $\pm 2$, $\pm 4$, $\pm 10$, $\pm 20$, are no longer on our list because we are only concerned with the possible factors of 5, rather than the possible factors of 20. In the process of factoring a polynomial, it is possible to whittle down the list because the constant term changes. We will never add to the list, though. The *initial* list of rational zeros that you create will always be the longest. Going back to the problem, we can factor the quadratic function that remains, and answer the question. The factored form of $P(x) = 2x^4 - x^3 - 42x^2 - 19x + 20$

is $(x + 1)(x + 4)(2x - 1)(x - 5)$. Notice that the rational zeros of this polynomial are -1, –4, 5, and $\dfrac{1}{2}$. All four of these numbers were on our original list.

---

When factoring polynomials, there are some observations that you can make to possibly shorten your workload. For example, consider the polynomial $2x^5 + 23x^4 + 89x^3 + 137x^2 + 89x + 20$. The process of synthetic division involves multiplication and addition, and our goal is to end up with 0 in the last column. The polynomial $2x^5 + 23x^4 + 89x^3 + 137x^2 + 89x + 20$ only has positive signs. If we put a positive number in the half-box, we will only be multiplying positive numbers together, and we will also be adding positive numbers together. As we progress through the synthetic division, our numbers will only get larger. The signs of the coefficients of the polynomials can guide us in our quest to factor the polynomial. When trying to factor $2x^5 + 23x^4 + 89x^3 + 137x^2 + 89x + 20$, I would not try any of the positive numbers in the original list of possible factors.

When looking for the zeros or the factors of a polynomial, any symmetry that the polynomial has may shorten your work. If $x = c$ is a zero of an even polynomial $P(x)$, then $P(c) = 0$. By symmetry, $P(-c) = P(c) = 0$. Similarly, if $x = c$ is a zero of an odd polynomial $P(x)$, then the symmetry of the polynomial means that $P(-c) = -P(c) = 0$. From this we see that the zeros of a symmetric polynomial will also be symmetric.

When a polynomial is written in factored form, the zeros of the polynomial can be easily determined by setting each factor equal to 0 and solving for $x$. In the complete factorization of a polynomial, some factors may appear more than once. If the factor $(x - c)$ occurs $k$ times in the complete factorization of a polynomial $P(x)$, then $c$ is called a root of $P(x)$ with multiplicity $k$. The function $P(x) = (x + 2)^3(x - 3)^2(x + 4)$ is a degree 6 polynomial. It has a zero of multiplicity 3 at $x = -2$, a zero of multiplicity 2 at $x = 3$, and a zero of multiplicity 1 at $x = -4$. Notice that the sum of all of the multiplicities of the zeros is 6, which is the degree of the polynomial. This will always happen.

## Lesson 4-4 Review

Factor the following polynomials:

1. $x^3 + x^2 - 8x - 12$
2. $2x^4 + 3x^3 - 9x^2 + x + 3$

# Lesson 4-5: Reconstructing Polynomials

Finding the zeros of a polynomial involves factoring the polynomial. In the previous lesson, we discussed a strategy for factoring polynomials that focused on the rational zeros of polynomials. Keep in mind that a degree n polynomial has n zeros. These zeros may or may not be distinct. The zeros will not be distinct if the polynomial has any repeating factors in its factorization. In this lesson, we will reconstruct a polynomial from its zeros.

Suppose that a degree 3 polynomial has zeros at $x = 1, x = 4$, and $x = -2$. Do we have enough information to construct the polynomial? Well, the function $P(x) = (x - 1)(x - 4)(x + 2)$ is a degree 3 polynomial with zeros at $x = 1, x = 4$, and $x = -2$. But so are the functions $Q(x) = 3(x - 1)(x - 4)(x + 2)$ and $T(x) = -4(x - 1)(x - 4)(x + 2)$. If we know all of the zeros of a polynomial (and their multiplicities), then we know a lot about the polynomial, but we cannot give a definitive formula for the polynomial. The zeros of a polynomial indicate the places where the graph of the polynomial crosses (or just touches) the $x$-axis. In order to reconstruct a polynomial from its zeros, we also need to know one point that the polynomial passes through that is off of the $x$-axis.

## Example 1

Find the equation of a degree 4 polynomial with zeros of multiplicity 2 at $x = -6$ and $x = 3$.

**Solution:** There are many such polynomials, but any polynomial that satisfies these conditions must be of the form $P(x) = a(x + 6)^2(x - 3)^2$, where $a$ is a constant. One specific polynomial is $P(x) = 2(x + 6)^2(x - 3)^2$.

## Example 2

Find the equation of the degree 4 polynomial with zeros of multiplicity 2 at $x = -6$ and $x = 3$ whose $y$-intercept is the point $(0, 30)$.

**Solution:** There are many degree 4 polynomials with those specific zeros, and every polynomial that satisfies these conditions must be of the form $P(x) = a(x + 6)^2(x - 3)^2$, where $a$ is a constant. Because we are given one point that the polynomial passes through that is off of the $x$-axis, we can solve for the constant $a$. Use the fact that $P(0) = 30$ to solve for $a$: $P(0) = a(0 + 6)^2(0 - 3)^2$

$$30 = 324a$$

$$a = \frac{30}{324} = \frac{5}{54}$$

The equation of the polynomial is $P(x) = \frac{5}{54}(x + 6)^2(x - 3)^2$.

When reconstructing a polynomial from its zeros, keep in mind that the factors of the polynomial are of the form $(x - c)$, where $c$ is a zero of the polynomial. Also, remember that the zeros of a polynomial do not have to be real. They can also be complex. When reconstructing a polynomial from its complex zeros, you must use the fact that complex zeros travel in (conjugate) pairs.

## Example 3

Find the equation of a degree 4 polynomial with zeros at $x = -3, x = 2$, and $x = 4 - i$.

**Solution:** At first glance, it may seem as if you have not been given enough zeros. A degree 4 polynomial has four zeros, and you have only been given three zeros. You must realize that there is a fourth zero lurking around: the complex conjugate of the complex zero! The fourth zero is located at $x = 4 + i$. Now that we have enough zeros, we can construct a polynomial that satisfies the given conditions: $P(x) = (x + 3)(x - 2)(x - (4 - i))(x - (4 + i))$. This polynomial is not in proper form. The complex numbers that appear in the two factors ruin the whole look of the formula. In this case, we need to combine those two factors to create a quadratic function that doesn't have any real zeros. To accomplish this, we need to rearrange some of the terms and use the fact that $(a + bi)(a - bi) = a^2 + b^2$:

$$P(x) = (x + 3)(x - 2)(x - (4 - i))(x - (4 + i))$$

Distribute the negative sign and group the real numbers together

$$P(x) = (x + 3)(x - 2)((x - 4) + i)((x - 4) - i))$$

Use the fact that $(a + bi)(a - bi) = a^2 + b^2$

$$P(x) = (x + 3)(x - 2)((x - 4)^2 + 1)$$

Expand $(x - 4)^2$ and combine terms

$$P(x) = (x + 3)(x - 2)(x^2 - 8x + 17)$$

One polynomial that satisfies the given conditions is

$$P(x) = (x + 3)(x - 2)(x^2 - 8x + 17).$$

---

It is important to understand the difference between "find *a* polynomial..." and "find *the* polynomial..." problems. In mathematics, the phrase "find a polynomial..." means that the answer may not be unique. The phrase, "find the polynomial..." means that there is one unique solution to the problem. It is important to distinguish between *a* and *the* in the instructions.

## Lesson 4-5 Review

1. Find the equation of a degree 3 polynomial with zeros at $x = 2$ and $x = 3 - 2i$.

2. Find the y-intercept of the degree 4 polynomial with zeros of multiplicity 2 at $x = 2$ and $x = -1$ that passes through the point $(3, 64)$.

## Lesson 4-6: Solving Polynomial Equations

Solving a polynomial equation involves gathering all of the terms of the polynomial on one side of the equation and having 0 on the other side. At that point, we can factor the polynomial and set each factor equal to 0.

## Example 1

Solve the equation $x^3 = 3x - 2$.

**Solution:** Gather all of the terms of the polynomial on one side of the equation. Our problem is equivalent to solving the equation $x^3 - 3x + 2 = 0$. Once we factor $x^3 - 3x + 2$, we will need to set each factor equal to 0 and solve for x. Using the Rational Zero Theorem, the only possible rational zeros of $x^3 - 3x + 2$ are ±1, ±2, and $x^3 - 3x + 2 = (x - 1)^2(x + 2)$. Setting each factor equal to 0, we see that $x = 1$ and $x = -2$ are zeros of the polynomial, and hence are solutions to our original equation. As a verification of our solution, we could graph $f(x) = x^3$ and $g(x) = 3x - 2$ and see where they intersect. The graphs of these two functions are shown in Figure 4.2. Notice that the two functions intersect at $x = 1$ and $x = -2$, though at $x = 1$ it appears that the graphs of the two functions just glance off of each other.

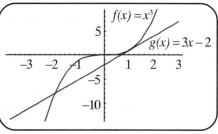

*Figure 4.2*

Now that we have a method for solving polynomial equations, we can solve a variety of equations that don't really involve polynomials, but can be transformed into polynomials. Recognizing these disguised polynomials takes some practice.

## Example 2

Solve the equation $x - 5\sqrt{x} + 6 = 0$.

**Solution:** There are two ways to approach this problem. One way is to understand that we can get rid of a square root by squaring. If we isolate the term that involves the radical, and then square both sides, the result will be a quadratic function. When using this technique, you must be careful to check your final answers. Whenever you square both sides of an equation, it is possible to introduce extraneous solutions. Once we square both sides of the equation, the result will be a quadratic function, which we can solve either by factoring, or by using the quadratic formula:

$$x - 5\sqrt{x} + 6 = 0$$

Move the $5\sqrt{x}$ term to the right side $\quad x + 6 = 5\sqrt{x}$

Square both sides of the equation $\quad (x+6)^2 = \left(5\sqrt{x}\right)^2$

Expand both sides of the equation $\quad x^2 + 12x + 36 = 25x$

Collect all of the terms on one side of the equation
$$x^2 - 13x + 36 = 0$$

Factor $\qquad\qquad\qquad\qquad (x-4)(x-9) = 0$

Set each factor equal to 0 and solve

The solutions to this equation are $x = 4$ or $x = 9$. Both of these numbers satisfy the original equation, so both solutions are valid.

---

The other approach to this problem involves making a substitution for $x$: if we let $z = \sqrt{x}$, then $z^2 = x$. We can then substitute into the original equation: $z^2 - 5z + 6 = 0$. We now have a quadratic equation that we can solve by factoring or by using the quadratic formula: $z^2 - 5z + 6 = (z-3)(z-2)$. Setting each factor equal to 0, we have: $z = 3$ or $z = 2$. This is all well and good, but the original problem asked us to solve for $x$, not for $z$. We need to substitute these values of $z$ into the equation $z^2 = x$ in order to find the values of $x$: $x = 3^2 = 9$ or $x = 2^2 = 4$. Either approach leads us to the same answer.

## Example 3

Solve the equation $x^{2/3} - x^{1/3} - 2 = 0$.

**Solution:** The fractional exponents are problematic. If the denominators of the powers of $x$ were 1, then we would have a quadratic equation. We can use the substitution technique that we used in Example 2 to solve this problem: let $z = x^{1/3}$. Then $z^3 = x$ and

$z^2 = \left(x^{1/3}\right)^2 = x^{2/3}$. Substituting into the original equation, we have: $z^2 - z - 2 = 0$.

This equation can be solved by factoring: $z^2 - z - 2 = (z-2)(z+1)$. Setting each factor equal to 0, we see that $z = 2$ or $z = -1$. We need to substitute these values of $z$ into the equation $z^3 = x$ in order to find the values of $x$: $x = 2^3 = 8$ or $x = (-1)^3 = -1$.

---

## Lesson 4-6 Review

Solve the following equations:

1. $x - 5\sqrt{x} - 6 = 0$

2. $x^{2/3} - 3x^{1/3} - 4 = 0$

# Lesson 4-7: Solving Polynomial Inequalities

A polynomial inequality occurs when a polynomial is less than 0, less than or equal to 0, greater than 0, or greater than or equal to 0. Solving a polynomial inequality is similar to solving a quadratic inequality, and we will apply the strategy we developed in the last chapter.

Because a polynomial is a continuous function, the zeros of a polynomial are the only place where a polynomial can change sign. Solving a polynomial inequality requires us to be able to find the zeros of the polynomial. If the polynomial is already factored, then it will be easy to find the zeros of the polynomial. Once we have all of the zeros of the polynomial, we can create a sign chart similar to the sign chart we developed for quadratic functions. We can then use the sign chart to solve the polynomial inequality. We will solve all polynomial inequalities using the same method. I will walk you through the solution to the quadratic inequality $(x - 2)(x + 2)(x - 4) > 0$ as I present the general procedure. In this lesson, I will work with polynomials that are already factored. You should realize that if you are working with a polynomial that is not factored, the first step is to make sure that 0 is on one side of the inequality, and then factor the polynomial.

1.  Make sure that 0 is on one side of the inequality. For the inequality $(x - 2)(x + 2)(x - 4) > 0$, this condition is met.

2.  Find the zeros of the polynomial that appears in the inequality. The first step in finding the zeros of a polynomial is to factor the polynomial. In our problem, the polynomial is already factored. Finding the zeros of the polynomial is a matter of setting each factor equal to 0 and solving for $x$:

    $(x - 2)(x + 2)(x - 4) = 0$

    $x = 2, x = -2,$ or $x = 4.$

3.  Arrange the zeros of the polynomial on a number line, as shown in Figure 4.3.

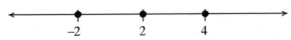

*Figure 4.3*

4.  The zeros of the polynomial break the number line into four regions. Select one test value from each of the four regions, and add these test values to the number line. Circle each test value, to distinguish them from the zeros of the polynomial. For our problem, I will choose the test values −3, 0, 3, and 5. Keep in mind that you can choose *any* number in each of the three regions.

5.  Evaluate the *sign* of the polynomial at each of these test values. There is no need to actually evaluate the function, just the sign of the function. I've outlined the logic of determining the sign of the function in the table below.

| $x$ | Sign of $(x-2)$ | Sign of $(x+2)$ | Sign of $(x-4)$ | Sign of $(x-2)$ $(x+2)(x-4)$ |
|---|---|---|---|---|
| −3 | − | − | − | − |
| 0 | − | + | − | + |
| 3 | + | + | − | − |
| 5 | + | + | + | + |

6.  Use the table to determine the sign of the polynomial in each of the three regions. The signs of the polynomial are shown in Figure 4.4.

*Figure 4.4*

7.  Answer the question. In this problem, we were asked to find the regions where $(x - 2)(x + 2)(x - 4) > 0$. From Figure 4.4, we can see that $(x - 2)(x + 2)(x - 4) > 0$ in the regions $(-2, 2) \cup (4, \infty)$.

   The solution to a polynomial inequality will include the zeros of the function if the inequality is ≥ or ≤. In this case, we would use brackets (either [ or ]) next to the zeros of the polynomial. If the inequality is strict (meaning > or <) we use parentheses next to the zeros of the polynomial.

## Example 1

Solve the polynomial inequality $(x - 3)^2(x - 1)(x + 1) \le 0$.

**Solution:** Follow the procedure that was previously outlined.

1.  Make sure that 0 appears on one side of the inequality.
2.  Find the zeros of the polynomial:

    $(x - 3)^2(x - 1)(x + 1) = 0$

    $x = 3, x = 1, \text{ or } x = -1$

3.  The number line is shown in Figure 4.5.

*Figure 4.5*

4.  Use the test values $x = -2, x = 0, x = 2$ and $x = 4$.
5.  The sign chart is shown in the following table:

| $x$ | Sign of $(x-3)^2$ | Sign of $(x-1)$ | Sign of $(x+1)$ | Sign of $(x-3)^2$ $(x-1)$ $(x+1)$ |
|---|---|---|---|---|
| $-2$ | $+$ | $-$ | $-$ | $+$ |
| $0$ | $+$ | $-$ | $+$ | $-$ |
| $2$ | $+$ | $+$ | $+$ | $+$ |
| $4$ | $+$ | $+$ | $+$ | $+$ |

6.  The signs of the polynomial are shown in Figure 4.6.

*Figure 4.6*

7.  From Figure 4.5, we can see that $(x - 3)^2(x - 1)(x + 1) \le 0$ in the region $[-1, 1]$.

## Lesson 4-7 Review

Solve the following inequalities:

1. $x(x + 2)(x - 4) \leq 0$

2. $-3(x + 2)(x - 1)^2(x - 4) < 0$

# Lesson 4-8: Graphing Polynomials

The graph of a constant function is a horizontal line. The graph of a polynomial with degree 1 is a line, and the graph of a quadratic function is a parabola. In this lesson, we will learn a strategy for graphing polynomials in general.

If the goal is to draw an accurate graph, then the tool to use is a graphing calculator. The problem with using a graphing calculator is that either you decide which part of the graph the calculator draws, or you rely on the calculator's standard viewing window. If you have a rough idea of how the polynomial looks, then you will be in a better position to set the calculator's parameters and graph the interesting regions of the polynomial.

When graphing a polynomial, don't be afraid to use the concepts we developed earlier. If a polynomial is even, then its graph will be symmetric about the $y$-axis, and an odd polynomial is symmetric across the origin. The sign of the leading coefficient and the degree of the polynomial will determine the behavior of the polynomial at extreme values of $x$. As $x$ takes on large positive values, we write $x \to \infty$, and as $x$ takes on large negative values, we write $x \to -\infty$. As $x \to \infty$ or $x \to -\infty$, a polynomial will also either head towards $+\infty$ or $-\infty$. The four possible situations are summarized in the table that follows.

| Degree of $P(x)$ | Sign of the leading coefficient | Behavior as $x \to -\infty$ | Behavior as $x \to \infty$ |
| --- | --- | --- | --- |
| Even | Positive | $P(x) \to \infty$ | $P(x) \to \infty$ |
| Even | Negative | $P(x) \to -\infty$ | $P(x) \to -\infty$ |
| Odd | Positive | $P(x) \to -\infty$ | $P(x) \to \infty$ |
| Odd | Negative | $P(x) \to \infty$ | $P(x) \to -\infty$ |

The graph of a polynomial will either start off large and negative, or large and positive. The graph of a polynomial will end by either going up towards $+\infty$ or shooting down towards $-\infty$.

The key to graphing a polynomial is to create a sign chart. Polynomials are continuous functions, so the entire graph should be one connected curve. It may have peaks and valleys, and it may or may not cross the $x$-axis. Every polynomial will have a $y$-intercept.

## Example 1

Figure 4.7 shows the sign chart for a polynomial. Graph the polynomial.

*Figure 4.7*

**Solution:** Create a set of co-ordinate axes and plot the zeros of the polynomial on the $x$-axis. As you read the sign chart from left to right, draw the polynomial above or below the $x$-axis, consistent with the sign chart. The graph of the polynomial is shown in Figure 4.8.

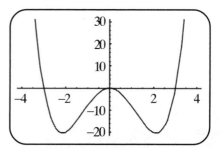

*Figure 4.8*

## Example 2

Graph the polynomial $P(x) = (x - 3)^2(x - 1)(x + 1)$.

**Solution:** We created a sign chart for this polynomial in Example 1 of Lesson 4-7, and the sign chart for this polynomial is shown again in Figure 4.9.

*Figure 4.9*

Based on this sign chart, Figure 4.10 is a sketch of the graph of this polynomial.

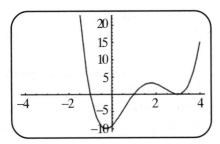

*Figure 4.10*

We can make some observations about the graphs of polynomials. Notice that when a factor of a polynomial has an even multiplicity, the graph of the polynomial just touches the $x$-axis and then turns around at the corresponding zero. Consider the polynomial that we graphed in Example 2. The multiplicity of the zero at $x = 3$ is 2 (an even number). The graph of the polynomial dips down to the $x$-axis and just touches it at $x = 3$, and then it changes direction. When a factor of a polynomial has an odd multiplicity, the graph of the polynomial crosses the $x$-axis and changes sign at the corresponding zero. Again, looking at the graph in Example 2, the graph of the polynomial crosses the $x$-axis at $x = 1$ and $x = -1$. This is an important observation, and it can be used to verify that any sketch of a polynomial is consistent with the factored form of the polynomial.

## Example 3

Can the graph shown in Figure 4.11 represent the graph of a polynomial of degree 3? Explain.

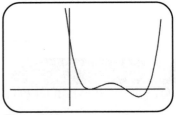

*Figure 4.11*

**Solution:** No. Notice that the graph crosses the $x$-axis in two places, and it just touches the $x$-axis in one place. The two zeros where the graph crosses the $x$-axis must have an odd multiplicity; the polynomial must have two factors of multiplicity greater than or equal to 1. The graph just touches the $x$-axis in one place, which means that the polynomial must have a factor of even multiplicity; the polynomial must have one factor with multiplicity greater than or equal to 2. If we add up the minimum multiplicities, we get $1 + 1 + 2 = 4$. In other words, the minimum degree of this polynomial is 4.

## Lesson 4-8 Review

1. Sketch the graph of the polynomial $P(x) = -3(x + 2)(x - 1)^2(x - 4)$.

2. What is the minimum degree of the polynomial shown in Figure 4.12?

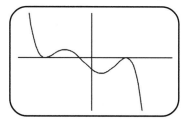

*Figure 4.12*

## Answer Key

### Lesson 4-1 Review

1. a. $f(x) = 3x^4 - 2x + 1$: neither

b. $f(x) = 3x^{10} - 2x^6 + 5$: even

c. $f(x) = 3x^9 - 2x^5 + 1$: neither

d. $f(x) = 3x^9 - 2x^5 + x$: odd

### Lesson 4-2 Review

Find the remainder of the following:

1. $\dfrac{x^{30} - 2x^5 + 3x - 1}{x + 1}$ : The remainder is $-1$.

2. $\dfrac{x^4 - 2x^2 - 8}{x^2 + 2}$ : The remainder is $0$.

3. The remainder of $\dfrac{P(x)}{x + 2}$ is $P(-2)$, which is $5$.

### Lesson 4-3 Review

1. Possible factors of 8: $\pm 1, \pm 2, \pm 4, \pm 8$, possible factors of 6: $\pm 1, \pm 2, \pm 3, \pm 6$, all possible combinations of factors of 8 divided by factors of 6:

$$\pm 1, \pm 2, \pm 4, \pm 8, \pm \tfrac{1}{2}, \cancel{\pm \tfrac{2}{2}}, \cancel{\pm \tfrac{4}{2}}, \cancel{\pm \tfrac{8}{2}}, \pm \tfrac{1}{3}, \pm \tfrac{2}{3}, \pm \tfrac{4}{3}, \pm \tfrac{8}{3}, \pm \tfrac{1}{6}, \cancel{\pm \tfrac{2}{6}}, \cancel{\pm \tfrac{4}{6}}, \cancel{\pm \tfrac{8}{6}}.$$

Eliminating duplicates, we have $\pm 1, \pm 2, \pm 4, \pm 8, \pm \tfrac{1}{2}, \pm \tfrac{1}{3}, \pm \tfrac{2}{3}, \pm \tfrac{4}{3}, \pm \tfrac{8}{3}, \pm \tfrac{1}{6}$.

2. No. If $(x - 7)$ is a factor of $6x^4 + 29x^3 - 24x^2 - 101x - 30$, that would mean that 7 is a factor of 30, which it is not.

## Lesson 4-4 Review

1. $x^3 + x^2 - 8x - 12 = (x+2)^2(x-3)$

2. $2x^4 + 3x^3 - 9x^2 + x + 3 = (x-1)^2(2x+1)(x+3)$

## Lesson 4-5 Review

1. $P(x) = (x-2)(x^2 - 6x + 13)$

2. The polynomial is $P(x) = 4(x-2)^2(x+1)^2$, and the $y$-intercept of this polynomial is $(0, 16)$.

## Lesson 4-6 Review

1. $x = 36$

2. $x = 1{,}024$ or $x = -1$

## Lesson 4-7 Review

1. $x(x+2)(x-4) \le 0$: $(-\infty, -2] \cup [0, 4]$

2. $-3(x+2)(x-1)^2(x-4) < 0$: $(-\infty, -2) \cup (4, \infty)$

## Lesson 4-8 Review

1. The graph of $P(x) = -3(x+2)(x-1)^2(x-4)$ is shown in Figure 4.13.

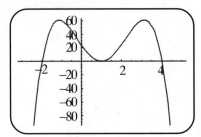

*Figure 4.13*

2. The minimum degree is 5 (there are two zeros with even multiplicity and one zero with odd multiplicity).

# Rational Functions

A *rational function* is a function that is the ratio of two polynomials. We will use our understanding of polynomials to analyze rational functions. Understanding the behavior of rational functions can help in making a variety of financial decisions. For example, energy efficient light bulbs last longer than incandescent bulbs, but they cost more. Analyzing the average cost of the two types of light bulbs may help you decide which type of light bulb to purchase. In this situation, the average cost of the light bulbs can be modeled using a rational function.

## Lesson 5-1: The Domain of a Rational Function

A rational function is a ratio of two polynomials. The domain of a rational function consists of all real numbers except the zeros of the polynomial in the denominator of the rational function.

### Example 1

Find the domain of the following rational functions. Write your answer in interval notation.

a. $\dfrac{x(x-1)}{(x+2)(x-3)}$

b. $\dfrac{x}{x^2+1}$

**Solution:** Set the denominator of each rational function equal to 0 and solve:

a. $(x + 2)(x - 3) = 0$ at $x = -2$ and $x = 3$, so the domain of the

function $\dfrac{x(x-1)}{(x+2)(x-3)}$ is $(-\infty, -2)\cup(-2, 3)\cup(3, \infty)$.

b. $x^2 + 1 \neq 0$ for all $x$, so the domain of the function $\dfrac{x}{x^2+1}$

is $(-\infty, \infty)$.

---

Let's compare the rational function $R(x) = \dfrac{x^2-4}{x-2}$ and the linear function $f(x) = x + 2$. These two functions are very similar to each other. Notice that $R(3) = 5$, and $f(x) = 5$. In fact, these functions agree, or have the same value, for almost every value of $x$ that you use. But there is one value of $x$ that gives different results: $x = 2$.

The numerator of this rational function can be factored:

$$R(x) = \frac{x^2 - 4}{x - 2} = \frac{(x-2)(x+2)}{x-2}.$$

At this point, we may be tempted to cancel the common factors and pretend they were never there. Unfortunately, we are not allowed to do this. The functions $R(x) = \dfrac{x^2-4}{x-2}$ and $f(x) = x + 2$ are different functions, specifically because the two functions have different domains. The domain of the rational function is $(-\infty, 2) \cup (2, \infty)$, and the domain of the linear function is $(-\infty, \infty)$. If someone takes the time to define a function as $R(x) = \dfrac{x^2-4}{x-2}$, then we must accept their definition as it is given. In this situation, we are not allowed to cancel these common factors; in doing so we are changing the domain of the function. Changing the domain of a function changes the function itself.

You may wonder what the graph of the function $R(x) = \dfrac{x^2-4}{x-2}$ looks like.

Well, because the function $R(x) = \dfrac{x^2-4}{x-2}$ gives the same numerical value as the linear function $f(x) = x + 2$ for all values of $x$ other than $x = 2$, the graph of $R(x) = \dfrac{x^2-4}{x-2}$

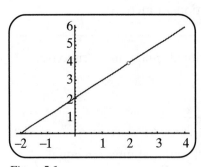

Figure 5.1

looks almost identical to the graph of $f(x) = x + 2$. The only difference between the two graphs is at the point $x = 2$. At $x = 2$, $R(x)$ is not defined, meaning that at $x = 2$, the graph of $R(x)$ has a little hole. This hole is so small that your graphing calculator might miss it. But when you sketch the graph of $R(x) = \frac{x^2-4}{x-2}$, you must show the little hole at $x = 2$. It is important that you realize that $R(x)$ is not defined at $x = 2$, and reflect this fact in your graph. The graph of $R(x) = \frac{x^2-4}{x-2}$ (with the little hole at $x = 2$) is shown in Figure 5.1.

## Lesson 5-1 Review

1. Find the domain of the following rational functions. Write your answers in interval notation.

   a. $\dfrac{2x-1}{x(x^2+4)}$

   b. $\dfrac{x-1}{x^2-1}$

2. Sketch the graph of the function $R(x) = \dfrac{x^2+x-2}{x-1}$.

# Lesson 5-2: Asymptotes

A rational function is a function of the form $f(x) = \frac{P(x)}{Q(x)}$ where $P(x)$ and $Q(x)$ are polynomials. One of the important features of a rational function is its **asymptotic behavior**. Asymptotic behavior is a mathematical term that refers to the behavior of a function as either $x$ or the function approaches infinity.

A rational function can have vertical, horizontal, or oblique asymptotes. A rational function will have a **vertical asymptote** if the magnitude of the function values becomes arbitrarily large, or tends to infinity, as $x$ approaches some fixed, finite number. A rational function will have a **horizontal asymptote** if the function approaches some fixed, finite number as the magnitude of $x$ gets arbitrarily large, or tends to infinity. A rational function will have an oblique asymptote if the function behaves as a linear function as $x$ gets arbitrarily large, or tends to infinity.

In order for a function to have a vertical or a horizontal asymptote, the magnitude of one of the variables, either the dependent or the independent variable, must head towards infinity while the other variable heads towards a finite number.

Let's look at the behavior of a specific function to help illustrate this idea. The domain of the rational function $f(x) = \frac{1}{x}$ is the set of all values of $x$ other than $x = 0$. The domain of $f(x) = \frac{1}{x}$ is $(-\infty, 0) \cup (0, \infty)$. This function has a vertical asymptote at $x = 0$. Notice that as $x$ gets close to 0, the magnitude of $f(x)$ becomes very large. The table here shows the value of $f(x)$ for very small positive values of $x$.

| x | f(x) |
|---|---|
| 0.00001 | 100,000 |
| 0.000001 | 1,000,000 |
| 0.0000001 | 10,000,000 |
| 0.00000001 | 100,000,000 |

We say that as $x$ gets close to 0, $f(x)$ gets very large. The reciprocal of a very small positive number will be a very large positive number. We can write this idea using shorthand notation. We can abbreviate the idea that $x$ is a positive number that gets close to 0 as $x \to 0^+$, and we can write the idea "the magnitude of $f(x)$ gets very large" as $|f(x)| \to \infty$. The little "+" above the 0 in $x \to 0^+$ just reminds us that $x$ is taking on *positive* values that are close to 0.

| x | f(x) |
|---|---|
| −0.00001 | −100,000 |
| −0.000001 | −1,000,000 |
| −0.0000001 | −10,000,000 |
| −0.00000001 | −100,000,000 |

Let's examine the function $f(x) = \frac{1}{x}$ for negative values of $x$ that are close to 0. The adjacent table shows the value of $f(x)$ for very small negative values of $x$.

The only differences between the two tables are the signs of the dependent and independent variables. When $x < 0$, $f(x) < 0$, but the *magnitude* of $f(x)$ is getting very large. In this situation, $x$ is taking on negative values that are close to 0, and $f(x)$ is heading towards $-\infty$. We can represent this more succinctly using the shorthand notation introduced earlier: "as $x \to 0^-$, $f(x) \to -\infty$." In both situations, $x$ is heading towards a finite number, namely 0, and the *magnitude* of $f(x)$ is heading towards infinity; the function $f(x) = \frac{1}{x}$ has a *vertical asymptote* at $x = 0$.

The vertical asymptotes of a rational function are fairly easy to recognize. Factor both the numerator and the denominator of the rational function, and find the zeros of the denominator (by setting each factor in the denominator equal to 0 and solving for $x$). The rational function will have a vertical asymptote at any value of $x$ that is a zero of the denominator and *not* a zero of the numerator. For the function $f(x) = \frac{1}{x}$, the denominator is 0 when $x = 0$, and the numerator is never $0$; $x = 0$ is a zero of the denominator and not a zero of the numerator. We see that $x = 0$ will be a vertical asymptote of the function $f(x) = \frac{1}{x}$.

The function $R(x) = \frac{x^2 - 4}{x - 2}$ does not have a vertical asymptote at $x = 2$. In this case, $x = 2$ is not in the domain of $R(x)$, but $x = 2$ is a zero of both the numerator and the denominator of $R(x)$. The graph of $R(x)$ does not shoot off to infinity as $x$ approaches 2.

To find the vertical asymptotes of a rational function, first find the zeros of the denominator. Then check to see if any of the zeros of the denominator are also zeros of the numerator. If they are, you will need to do some more exploring. Remember that a vertical asymptote is characterized by the numerator being finite while the denominator heads towards 0.

To find the horizontal asymptote of a rational function, we need to understand the nature of a polynomial as the magnitude of the independent variable becomes large. As $x$ becomes large, only the leading term becomes important, and dominates the behavior of the polynomial. As $x$ becomes large, the polynomial $f(x) = 3x^4 - 2x^3 + 100$ behaves like $3x^4$. For really large values of $x$, the terms $-2x^3$ and 100 have a negligible effect on the value of the function, and they can be ignored. Notice that this simplification only holds for large values of $x$. If $x$ is not extremely large, then the other terms in the polynomial are important. Fortunately, the horizontal asymptotes of a function will be found when the independent variable is very large in magnitude.

A rational function is a ratio of two polynomials. The first step in finding the horizontal asymptote of a rational function is to focus on the leading term of the two polynomials that make up the rational function. If $R(x)$ is a rational function, we can write $R(x)$ as a ratio of two polynomials:

$$R(x) = \frac{a_n x^n + a_{n-1} x^{n-1} + \ldots + a_1 x + a_0}{b_m x^m + b_{m-1} x^{m-1} + \ldots + b_1 x + b_0}.$$

When $x$ is large, $R(x) \sim \frac{a_n x^n}{b_m x^m}$. I am using the symbol $\sim$ to mean "behaves like." There are three cases that we need to consider.

1. The degree of the numerator is greater than the degree of the denominator, or $n > m$. In this case, we can cancel some of

the powers of $x$ and $R(x) \sim \frac{a_n}{b_m} x^{n-m}$, with $n - m > 0$. As $x$ gets large, the magnitude of this rational function also gets large, so there is no horizontal asymptote. Remember that a horizontal asymptote will occur when the function values approach a finite number as the independent variable gets large. If the degree of the numerator of a rational function is larger than the degree of the denominator, the rational function will not have a horizontal asymptote.

2. The degree of the numerator equals the degree of the denominator, or $n = m$. In this case, we can cancel all of the powers of $x$ and $R(x) \sim \frac{a_n}{b_n}$. In other words, the rational function gets close to the value of the ratio of the leading coefficients. The horizontal asymptote will be the line $y = \frac{a_n}{b_n}$.

3. The degree of the numerator is less than the degree of the denominator, or $n < m$. In this case, we can cancel out some of the powers of $x$ and $R(x) \sim \frac{a_n}{b_m x^{m-n}}$, with $m - n > 0$. As $x$ gets large, the value of $R(x)$ heads towards 0, and the horizontal asymptote will be the line $y = 0$.

A rational function will have, at most, one horizontal asymptote.

The intercepts of a rational function are also worth finding. The $y$-intercept of a rational function is found by evaluating the function at $x = 0$. Of course, if $x = 0$ is not in the domain of the rational function (because the function has a vertical asymptote at $x = 0$), then the rational function will not have a $y$-intercept. The $x$-intercepts of a rational function are simply the zeros of the numerator. If the numerator and denominator of a rational function have already been factored (to find the vertical asymptotes of the function), then most of the work necessary to find the $x$-intercepts has already been done. This will be illustrated in Example 1.

### Example 1

Find the intercepts and the asymptotes of the following rational functions:

a. $R(x) = \dfrac{x^2 - 2x + 1}{2x^2 + 2x - 12}$

b. $F(x) = \dfrac{(x-2)(x+3)}{2(x+1)^2(x-1)}$

**Solution:**

a. To find the *y*-intercept of $R(x) = \dfrac{x^2 - 2x + 1}{2x^2 + 2x - 12}$, evaluate $R(0)$:

$$R(0) = \frac{0^2 - 2 \cdot 0 + 1}{2 \cdot 0^2 + 2 \cdot 0 - 12} = -\frac{1}{12}.$$

The *y*-intercept is the point $\left(0, -\dfrac{1}{12}\right)$. To find the *x*-intercepts and the vertical asymptotes, factor the polynomials in the numerator and the denominator:

$$R(x) = \frac{(x-1)^2}{2(x+3)(x-2)}.$$

The zeros of the denominator are $x = -3$ and $x = 2$, neither of which are zeros of the numerator. Therefore the vertical asymptotes of the function $R(x)$ are $x = -3$ and $x = 2$. The *x*-intercepts are found by finding the zeros of the numerator; there is only one *x*-intercept: (1, 0). To find the horizontal asymptote, notice that the degree of the numerator and the

degree of the denominator are the same, so $R(x) \sim \dfrac{x^2}{2x^2} = \dfrac{1}{2}$.

Therefore, the horizontal asymptote is the line $y = \dfrac{1}{2}$.

b. To find the *y*-intercept of $F(x) = \dfrac{(x-2)(x+3)}{2(x+1)^2(x-1)}$, evaluate $F(0)$:

$$F(0) = \frac{(0-2)(0+3)}{2(0+1)^2(0-1)} = \frac{-6}{-2} = 3.$$

The *y*-intercept is the point (0, 3). The polynomials in the numerator and the denominator are already factored, so most of the work in finding the *x*-intercepts and the asymptotes has already been done. The *x*-intercepts are the points (2, 0) and (−3, 0). The vertical asymptotes are the zeros of the denominator that are not zeros of the numerator. The zeros of the denominator of $F(x)$ are $x = -1$ and $x = 1$, neither of which are zeros of the numerator. Therefore the vertical asymptotes of $F(x)$ are $x = -1$ and $x = 1$. To find the horizontal asymptote, notice that the degree of the numerator is 2 and the degree of the denominator is 3, so the horizontal asymptote is the line $y = 0$.

A rational function will have an oblique asymptote if the degree of the numerator is one more than the degree of the denominator. To find the oblique asymptote, divide the polynomial in the numerator by the polynomial in the denominator. The quotient will be a linear function, and the remainder will be a new rational function whose numerator has a smaller degree than its denominator. The oblique asymptote will be the quotient, or the linear function.

## Example 2

Find the oblique asymptote of the function $R(x) = \dfrac{2x^2 - 3x + 4}{x - 1}$.

**Solution:** Use synthetic division to find the quotient and the remainder: $R(x) = \dfrac{2x^2 - 3x + 4}{x - 1} = 2x - 1 + \dfrac{3}{x - 1}$.

The oblique asymptote is $y = 2x - 1$.

## Lesson 5-2 Review

Find the intercepts and the asymptotes of the following functions:

1. $G(x) = \dfrac{x^2 + x - 6}{2x^2 - x - 3}$    2. $H(x) = \dfrac{x^2 - 1}{x^3 - 5x^2 + 6x}$    3. $F(x) = \dfrac{x^2 - 3x - 4}{x + 2}$

# Lesson 5-3: Rational Inequalities

A *rational inequality* occurs when a rational function is less than 0, less than or equal to 0, greater than 0, or greater than or equal to 0. Solving a rational inequality is similar to solving a polynomial inequality, and we will solve rational inequalities using a technique similar to the one we used to solve polynomial inequalities.

A rational function has both zeros and discontinuities, or places where the denominator is equal to zero. These are the only places where a rational function can change sign. Solving a rational inequality requires us to be able to find the zeros and the discontinuities of the rational function. If the rational function is already factored, then it will be easy to find the zeros of the rational function: set each factor of the polynomial in the numerator equal to zero. To find the discontinuities of a rational function, set each factor of the polynomial in the denominator equal to zero. Once we have all of the zeros and discontinuities of the rational function, we can create a sign chart similar to the sign chart we developed for polynomials. We can then use the sign chart to solve the rational inequality. We will solve all rational inequalities using the same method. I will walk you through the solution to the quadratic inequality $\frac{(x-2)(x+2)}{x(x-4)} \leq 0$ as I present the general procedure. In this lesson, I will work with rational functions that are already factored. You should realize that if you are working with a rational function that is not factored, the first step is to make sure that 0 is on one side of the inequality, and then factor the rational function.

1.  Make sure that 0 is on one side of the inequality. For the inequality $\frac{(x-2)(x+2)}{x(x-4)} \leq 0$, this condition is met.

2.  Find the zeros of the rational function. The first step in finding the zeros of a rational function is to factor the polynomial in the numerator. In our problem, the polynomial in the numerator is already factored. Finding the zeros of the rational function is a matter of setting each factor equal to 0 and solving for $x$:

    $(x - 2)(x + 2) = 0$
    $x = 2$ or $x = -2$

3.  Find the discontinuities of the rational function. The first step in finding the discontinuities of a rational function is to factor the polynomial in the denominator. In our problem, the polynomial in the denominator is already factored. Finding the discontinuities of the rational function is a matter of setting each factor in the denominator equal to 0 and solving for x:

    $x(x - 4) = 0$
    $x = 0$ or $x = 4$

4. Arrange the zeros and the discontinuities of the rational function on a number line, as shown in Figure 5.2.

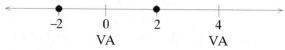

*Figure 5.2*

5. The zeros and the discontinuities of the rational function break the number line into five regions. Select one test value from each of the five regions, and add these test values to the number line. Circle each test value, to distinguish them from the zeros or discontinuities of the rational function. For our problem, I will choose the test values −3, −1, 1, 3, and 5. Keep in mind that you can choose *any* number in each of the three regions.

6. Evaluate the *sign* of the polynomial at each of these test values. There is no need to actually evaluate the function, just the sign of the function. I've outlined the logic of determining the sign of the function in the table shown here.

| $x$ | Sign of $x$ | Sign of $(x-2)$ | Sign of $(x+2)$ | Sign of $(x-4)$ | Sign of $\dfrac{(x-2)(x+2)}{x(x-4)}$ |
|---|---|---|---|---|---|
| −3 | − | − | − | − | + |
| −1 | − | − | + | − | − |
| 1 | + | − | + | − | + |
| 3 | + | + | + | − | − |
| 5 | + | + | + | + | + |

7. Use the table to determine the sign of the rational function in each of the three regions. The signs of the rational function are shown in Figure 5.3.

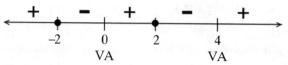

*Figure 5.3*

8. Answer the question. In this problem, we were asked to find the regions where $\frac{(x-2)(x+2)}{x(x-4)} \leq 0$. From Figure 5.3, we can see that $\frac{(x-2)(x+2)}{x(x-4)} \leq 0$ in the regions $[-2, 0) \cup [2, 4)$.

The solution to a rational inequality will include the zeros of the function if the inequality is $\geq$ or $\leq$. The discontinuities of a rational function will never be included, because those points are not in the domain of the function. If the inequality is strict (meaning $>$ or $<$), do not include the zeros of the rational function; use parentheses next to the zeros of the rational function.

## Lesson 5-3 Review

Solve the following inequalities:

1. $\dfrac{(x-3)^2}{(x-1)(x+1)} < 0$

2. $\dfrac{-3(x+2)}{(x-1)^2} \geq 0$

# Lesson 5-4: Graphing Rational Functions

In Chapter 4, we developed a strategy for graphing polynomials. We can modify that strategy to enable us to graph rational functions. This is another example of how mathematics builds on concepts that were previously developed.

As I discussed with polynomials, if the goal is to draw an accurate graph, then the tool to use is a graphing calculator. Remember that the problem with using a graphing calculator is that either you decide which part of the graph the calculator draws, or you rely on the calculator's standard viewing window. If you have a rough idea of how the rational function looks, then you will be in a better position to set the calculator's parameters and graph the interesting regions of the rational function. Also, the typical settings for the viewing window may not indicate the asymptotic behavior of the rational function. Being able to analyze the graph of a rational function algebraically will usually provide more insight into the function than just looking at the calculator screen.

When graphing a rational function, it is important that you make use of the ideas that we developed in the previous lessons. If a rational function is even, then its graph will be symmetric about the y-axis, and an odd rational function is symmetric across the origin. If a rational function has a horizontal asymptote, then you know the behavior of the rational function at extreme values of $x$: the horizontal asymptote indicates where the graph of the rational function should start and end.

The key to graphing a rational function is to create a sign chart. A rational function will be continuous between its discontinuities. The discontinuities of a rational function are either vertical asymptotes or holes. It is easy to recognize a hole. A hole will occur if both the numerator and the denominator share a common factor, and the multiplicity of the factor in the numerator is greater than or equal to the multiplicity of the factor in the denominator. For example, for the function $f(x) = \frac{x^2-1}{x+1}$, both the numerator and the denominator share a common factor: $(x + 1)$. The multiplicity of this factor is 1 for both the numerator and the denominator. The point $x = -1$ is a discontinuity of $f(x)$, but it will be a hole rather than a vertical asymptote. For the function $g(x) = \frac{(x-1)^2(x+1)}{(x-1)}$, both the numerator and the denominator share a common factor: $(x = -1)$. The multiplicity of this factor is 2 in the numerator and 1 in the denominator. The point $x = 1$ is a discontinuity of $g(x)$, but it will be a hole.

A vertical asymptote will occur if the denominator has any zeros that are not also zeros of the numerator, or if the numerator and the denominator share a common factor, and the multiplicity of the factor in the numerator is less than the multiplicity of the factor in the denominator. For example, for the function $f(x) = \frac{1}{x-1}$, $(x - 1)$ is a factor of the numerator, but not the denominator. This function will have a vertical asymptote at $x = 1$. For the function $g(x) = \frac{x^2-1}{(x-1)^2}$, both the numerator and the denominator share a common factor: $(x - 1)$. The multiplicity of this factor is 1 in the numerator and 2 in the denominator. This function will have a vertical asymptote at $x = 1$.

Once you find all of the zeros of a rational function (by setting the polynomial in the numerator equal to 0 and solving), and all of the discontinuities of a rational function (by setting the polynomial in the denominator equal to 0 and solving), you are ready to create a sign chart for the rational function. After you create the sign chart for the rational function, determine if the function has a horizontal asymptote. Based on the signs, the zeros, vertical asymptotes, and the horizontal asymptote of the function, you can sketch a graph of the rational function. Realize that the sketch you draw will not be exact. The rational function may have peaks and dips that you didn't draw. There are techniques that are discussed in calculus that will enable you to more accurately draw the graph of a rational function.

## Example 1

Figure 5.4 shows the sign chart for a rational function. Suppose that the rational function has zeros at $x = -2$ and $x = 3$, and a vertical asymptote at $x = 0$. Also, suppose that $y = 1$ is a horizontal asymptote. Sketch a graph of the rational function.

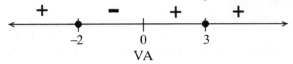

*Figure 5.4*

**Solution:** The function has a vertical asymptote at $x = 0$, so we will have to graph this function in two stages. The first stage involves the graph of the function for values of $x$ between $-\infty$ and $0$, and the second stage deals with values of $x$ between $0$ and $\infty$. Because $y = 1$ is a horizontal asymptote of the rational function, we know that the graph of the function has to start and end near the line $y = 1$. The function starts out near $y = 1$, and the first thing it has to do is cross the $x$-axis at the point $(-2, 0)$. From the sign chart of the function, we see that the function changes sign at $(-2, 0)$ and becomes negative. The next important aspect of this function is that there is a vertical asymptote at $x = 0$. A vertical asymptote acts like a brick wall: functions are not allowed to cross a vertical asymptote. The vertical asymptote will guide the function as it heads towards $-\infty$. That takes care of the first stage of the graph. For the second stage, our function is positive on the right side of the vertical asymptote, so the function will be coming from $+\infty$. The function heads down to touch the $x$-axis at $(3, 0)$, and then it will go back up and head towards the horizontal asymptote. Remember that we are drawing rough sketches of the function. Your graph doesn't have to look identical to mine. Your graph does have to start at the horizontal asymptote, cross the $x$-axis at $(-2, 0)$ and head towards $-\infty$ along the vertical asymptote $x = 0$. Your graph must also come down from $+\infty$, touch the $x$-axis, and head

towards the horizontal asymptote. We don't have the tools to
locate the peaks and valleys of
the graph. All we can do is determine where to start and end
the graph, where to cross or
touch the x-axis, and where to
head towards infinity. My graph
is shown in Figure 5.5.

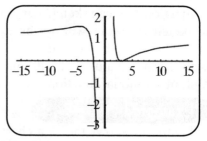

Figure 5.5

## Example 2

Figure 5.6 shows the sign chart for a rational function. Suppose
that the rational function has a zero at $x = -1$, and vertical
asymptotes at $x = -4$ and $x = 3$. Also, suppose that $y = 0$ is a
horizontal asymptote. Sketch a graph of the rational function.

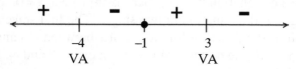

Figure 5.6

**Solution:** The function has vertical asymptotes at $x = -4$ and $x = 3$
so we will have to graph this function in three stages. The first
stage involves the graph of the function for values of $x$ between $-\infty$
and $-4$, the second stage deals with values of $x$ between $-4$ and $3$,
and the third stage deals with values of $x$ between 3 and $\infty$. Because $y = 0$ is a horizontal asymptote of the rational function, we
know that the graph of the function has to start and end near the
line $y = 0$. The function starts out near $y = 0$, and is positive according to the sign chart. The first thing the function has to do is
head towards $+\infty$ along the vertical asymptote $x = -4$. For the next
stage, the rational function is negative, so the function comes up
from $-\infty$. The function crosses the x-axis at the point $(-1, 0)$ and
changes sign to become positive. After that, the function heads
towards $+\infty$ along the vertical asymptote $x = 3$. For the third, and
final, stage, our function is negative on the right side of the vertical asymptote, so the function will be coming from $-\infty$. The function has nothing else to do (there are no more zeros and no more

vertical asymptotes), so it may as well just cruise into its horizontal asymptote. Remember that we are drawing rough sketches of the function. Your graph doesn't have to look identical to mine, but it must cross the *x*-axis at the right place, cruise to infinity along the vertical asymptotes, and

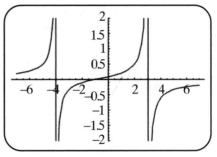

*Figure 5.7*

slide into the horizontal asymptote when you're done. My graph is shown in Figure 5.7.

---

## Lesson 5-4 Review

1.  Sketch a graph of $\dfrac{(x-3)^2}{(x-1)(x+1)}$.

2.  Figure 5.8 shows the sign chart for a rational function. Suppose that the rational function has a zero at $x = 0$, and vertical asymptotes at $x = -3$ and $x = 3$. Also, suppose that $y = 0$ is a horizontal asymptote. Sketch a graph of the rational function.

*Figure 5.8*

# Lesson 5-5: Partial Fraction Decomposition

The process of adding two rational functions involves finding a common denominator and then adding the numerators. There are times when it is beneficial to combine rational functions, and there other times when it is better to separate rational functions that have already been combined. The following procedure, called partial fraction decomposition, is used in calculus. Partial fraction decomposition is the reverse process of adding rational functions.

The first step in finding the partial fraction decomposition of a rational function is to make sure that the degree of the polynomial in the numerator is less than the degree of the polynomial in the denominator. If it isn't, you must perform the long division and decompose the remainder after the division. The technique I am about to present to you only works if the degree of the numerator is less than the degree of the denominator!

It is important to keep in mind that the motivation behind partial fraction decomposition involves reversing the process of adding fractions. An important theorem in algebra states that any polynomial $P(x)$ can be factored as a product of linear factors of the form $(ax + b)$ and irreducible quadratic factors of the form $ax^2 + bx + c$. An **irreducible quadratic factor** is a quadratic expression that has no real zeros. The expression $x^2 + 1$ is an example of an irreducible quadratic factor. To decompose a rational function $\frac{P(x)}{Q(x)}$, first factor the polynomial in the denominator, $Q(x)$ completely. Write the rational function $\frac{P(x)}{Q(x)}$ as a sum of fractions of the form $\frac{A}{(ax+b)^i}$ and $\frac{Ax+B}{(ax^2+bx+c)^i}$. The fractions to include depend on the nature of the factors of $Q(x)$. I will illustrate the various situations that can occur with examples.

Suppose the factors of $Q(x)$ are all distinct and linear:

$$(a_1x+b_1), (a_2x+b_2), ..., (a_nx+b_n).$$

Write $\frac{P(x)}{Q(x)} = \frac{A_1}{(a_1x+b_1)} + \frac{A_2}{(a_2x+b_2)} + ... \frac{A_n}{(a_nx+b_n)}$ and solve for the constants $A_1, A_2, ..., A_n$.

## Example 1

Find the partial fraction decomposition of $\frac{1}{(x-2)(x+3)}$.

**Solution:** The polynomial in the denominator is already factored. Decompose the rational function into individual fractions:

$$\frac{1}{(x-2)(x+3)} = \frac{A_1}{(x-2)} + \frac{A_2}{(x+3)}.$$

Solve for the constants $A_1$ and $A_2$. First, multiply both sides of the equation by $(x-2)(x+3)$:

$$1 = A_1(x+3) + A_2(x-2).$$

This equation must be true for all values of $x$. In particular, this equation must be true if $x = -3$. Substitute $x = -3$ into the equation $1 = A_1(x+3) + A_2(x-2)$ and solve for $A_2$:

$$1 = A_1(-3+3) + A_2(-3-2)$$

$$A_2 = -\frac{1}{5}$$

Substitute $x = 2$ into the equation $1 = A_1(x+3) + A_2(x-2)$ and solve for $A_1$:

$$1 = A_1(2+3) + A_2(2-2)$$

$$A_1 = \frac{1}{5}$$

So $\dfrac{1}{(x-2)(x+3)} = \dfrac{\frac{1}{5}}{(x-2)} + \dfrac{-\frac{1}{5}}{(x+3)}.$

---

Suppose the factors of $Q(x)$ are all linear, but some factors repeat. Suppose the factor $(a_1x + b_1)$ is repeated $r$ times. Then the linear factor $(a_1x + b_1)$ will be repeated $r$ times, in the form:

$$\frac{A_1}{(a_1x+b_1)} + \frac{A_2}{(a_2x+b_2)^2} + \cdots \frac{A_r}{(a_nx+b_n)^r}.$$

We would need to solve for the constants $A_1, A_2, \ldots, A_r$. The non-repeating factors are handled as before.

### Example 2

Decompose the function $\dfrac{x^2+1}{x(x-1)^3}$ into its components, but do not solve for the constants.

**Solution:** There is one linear factor that repeats 3 times, so we have:

$$\frac{x^2+1}{x(x-1)^3} = \frac{A_1}{x} + \frac{A_2}{(x-1)} + \frac{A_3}{(x-1)^2} + \frac{A_4}{(x-1)^3}.$$

---

If the factors of $Q(x)$ include a quadratic expression that has no real zeros, then the numerator of the fraction that incorporates the quadratic expression must be of the form $Bx + C$. Repeating factors are handled in the same way we discussed earlier. For example, the partial fraction

decomposition of $\frac{1}{x(x^2+1)}$ is: $\frac{1}{x(x^2+1)} = \frac{A_1}{x} + \frac{B_1 x + C_1}{(x^2+1)}$,

and the partial fraction decomposition of

$\frac{1}{x^3(x^2+1)^2}$ is: $\frac{1}{x^3(x^2+1)^2} = \frac{A_1}{x} + \frac{A_2}{x^2} + \frac{A_3}{x^3} + \frac{B_1 x + C_1}{(x^2+1)} + \frac{B_2 x + C_2}{(x^2+1)^2}$.

Finding the partial fraction decomposition of a rational function involves solving several equations for several variables. To help us solve these types of problems, we will need to develop a strategy for solving systems of equations. We will do this in Chapter 8.

## Lesson 5-5 Review

Find the partial fraction decomposition of the following functions:

1. $\dfrac{2x-2}{x^2-2x}$

2. $\dfrac{2x+8}{x^2-4}$

## Answer Key

### Lesson 5-1 Review

1. a. $\dfrac{2x-1}{x(x^2+4)} : (-\infty, 0) \cup (0, \infty)$

   b. $\dfrac{x-1}{x^2-1} : (-\infty, -1) \cup (-1, 1) \cup (1, \infty)$

2. The graph of $R(x) = \dfrac{x^2+x-2}{x-1}$ :

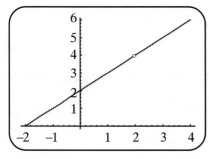

Figure 5.9

## Lesson 5-2 Review

1. $G(x) = \frac{x^2+x-6}{2x^2-x-3}$ :

   $x$-intercepts $(-3,0)$ and $(2,0)$,
   $y$-intercept $(0,2)$,

   vertical asymptotes $x = \frac{3}{2}$ and $x = -1$,

   horizontal asymptote $y = \frac{1}{2}$.

2. $H(x) = \frac{x^2-1}{x^3-5x^2+6x}$ :

   $x$-intercepts $(-1,0)$ and $(1,0)$,
   there is no $y$-intercept,
   vertical asymptotes $x = 0, x = 3$ and $x = 2$,
   horizontal asymptote $y = 0$.

3. $F(x) = \frac{x^2-3x-4}{x+2}$ :

   $x$-intercepts $(-1,0)$ and $(4,0)$,
   $y$-intercept $(0,-2)$,
   vertical asymptote $x = -2$,
   oblique asymptote $y = x - 5$.

## Lesson 5-3 Review

1. $\frac{(x-3)^2}{(x-1)(x+1)} < 0$ on $(-1,1)$     2. $\frac{-3(x+2)}{(x-1)^2} \geq 0$ on $(-\infty, -2]$

## Lesson 5-4 Review

1. The graph of $\frac{(x-3)^2}{(x-1)(x+1)}$ :

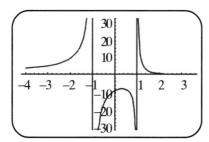

*Figure 5.10*

2.

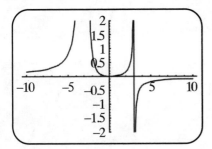

*Figure 5.11*

## Lesson 5-5

1. $\dfrac{2x-2}{x^2-2x} = \dfrac{1}{x-2} + \dfrac{1}{x}$

2. $\dfrac{2x+8}{x^2-4} = \dfrac{3}{x-2} - \dfrac{1}{x+2}$

# Power Functions

A power function is a function in which the dependent variable is proportional to a power of the independent variable. Power functions can be used to find the area of a circle ($A = \pi r^2$), or the volume of a sphere ($V = \frac{4}{3}\pi r^3$). Power functions can also be used to model the gravitational force between two objects. Power functions are used throughout physics and chemistry.

A **power function** is a function of the form $f(x) = ax^p$, where $a$ and $p$ are real numbers. The domain of a power function depends on the constant $p$. The power with $p = 0$ is the function: $f(x) = ax^0 = a$. The graph of this function is a horizontal line. In this chapter, we will examine power functions for various values of $p$. We will begin with positive integer values of $p$.

## Lesson 6-1: Positive Integer Powers

If $f(x) = ax^p$ where $p \geq 0$, then $x = 0$ will be in the domain of the function. The functions $f(x) = 2x^3$ and $g(x) = -3x^5$ are examples of power functions whose power is a positive integer. With both of these functions, $x = 0$ is in the domain, and both of these functions pass through the origin. If $p$ is a *non-negative* integer (meaning that $p$ is an integer that is *greater than or equal to 0*), then the domain of the function will be the set of all real numbers.

We will first consider power functions of the form $f(x) = x^p$ for positive values of $p$ that are odd. Figure 6.1 on page 126 shows the graphs of the functions $f(x) = x$, $f(x) = x^3$, and $f(x) = x^5$ on the same axes, for values of $x$ between −2 and 2.

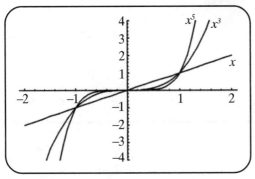

Figure 6.1

Notice that these graphs all intersect at the origin and at the points (1, 1) and (–1, –1). For values of $x$ in the interval (0, 1), we notice that the higher the power, the lower the graph of the function. For values of $x$ greater than 1, the opposite trend holds: higher powered functions are higher. For values of $x$ in the interval (–1, 0), the higher powered functions are higher, and for values of $x$ less than –1, the higher powered functions are lower. Observe that these functions are all odd, so they are symmetric about the origin. The concavity of these functions changes at $x = 0$: when $x < 0$ the functions are concave down, and when $x > 0$ these functions are concave up. The point where a function changes concavity is called an **inflection point.**

We will now consider power functions of the form $f(x) = x^p$ for positive values of $p$ that are even. Figure 6.2 shows the graphs of the functions $f(x) = x^2$, $f(x) = x^4$, and $f(x) = x^6$ on the same axes, for values of $x$ between –2 and 2.

Notice that these graphs all intersect at the origin and at the points (1, 1) and (–1, 1). For values of $x$ in the interval (0, 1), we notice that the higher the power, the lower the graph of the function. For values of $x$ greater than 1, the opposite trend holds: Higher powered functions are higher. For values of $x$ in the interval (–1, 0), the higher powered functions are

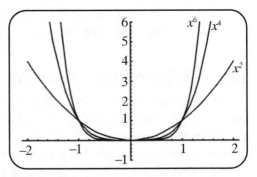

Figure 6.2

lower, and for values of $x$ less than –1, the higher powered functions are higher. Observe that these functions are all even, so they are symmetric about the $y$-axis. Notice that these functions are concave up throughout their domain.

# Lesson 6-2: Negative Integer Powers

We will now consider power functions of the form $f(x) = x^p$ for negative values of $p$ that are odd integers. If $p < 0$, then $x = 0$ will not be in the domain of the function. If $p$ is a negative integer, then the domain of the function will be all non-zero real numbers, or $(-\infty, 0) \cup (0, \infty)$. Figure 6.3 shows the graphs of the functions $f(x) = x^{-1}$, $f(x) = x^{-3}$, and $f(x) = x^{-5}$ on the same axes, for values of $x$ between $-2$ and $2$.

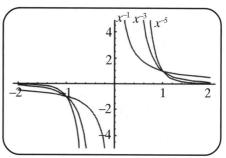

Notice that these functions are not defined at $x = 0$. The domain of these functions is the set of all non-

*Figure 6.3*

zero real numbers. These graphs intersect at the points $(1, 1)$ and $(-1, -1)$. For values of $x$ in the interval $(0, 1)$, we notice that the higher the magnitude of the power, the higher the graph of the function. For values of $x$ greater than 1, the opposite trend holds: functions whose powers are larger in magnitude are lower. For values of $x$ in the interval $(-1, 0)$, the larger the magnitude of the power, the lower the function, and for values of $x$ less than $-1$, the larger the magnitude of the power, the higher the function.

Observe that these functions are all odd, so they are symmetric about the origin. These functions are concave down when $x < 0$ and they are concave up when $x > 0$. These functions do not have an inflection point, however, because there is no point *on the graph* where the function changes concavity: the graph has a break when $x = 0$.

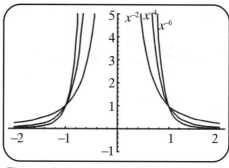

*Figure 6.4*

Next, let us turn our attention to power functions of the form $f(x) = x^p$ for negative values of $p$ that are even numbers. Figure 6.4 shows the graphs of the functions $f(x) = x^{-2}$, $f(x) = x^{-4}$, and $f(x) = x^{-6}$ on the same axes, for values of $x$ between $-2$ and $2$.

Notice that these graphs are not defined at $x = 0$. The domain of these functions is the set of all non-zero real numbers. These graphs intersect at the points $(1, 1)$ and $(-1, 1)$. For values of $x$ in the interval $(0, 1)$, we notice that the higher the magnitude of the power, the higher the graph of the function. For values of $x$ greater than 1, the opposite trend holds: functions whose powers are larger in magnitude are lower. For values of $x$ in the interval $(-1, 0)$, the larger the magnitude of the power, the higher the function, and for values of $x$ less than $-1$, the larger the magnitude of the power, the lower the function. Observe that these functions are all even, so they are symmetric about the $y$-axis. These functions are concave up for both negative and positive values of $x$.

## Lesson 6-3: Rational Powers

The next group of power functions to study involves positive rational powers. The function $f(x) = x^p$ is called the $p^{\text{th}}$ **power** of $x$, and $g(x) = x^{1/p}$ is called the $p^{\text{th}}$ **root** of $x$. Because some fractional powers involve roots that are only defined for non-negative values of $x$, we will analyze these functions for $x \geq 0$. Figure 6.5 shows the graphs of the functions $f(x) = x^2$, $f(x) = x^{3/2}$, $f(x) = x$, $f(x) = x^{1/2} >$, and $f(x) = x^{1/3}$.

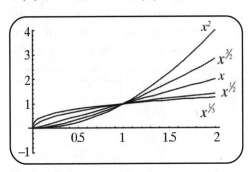

*Figure 6.5*

All of these functions pass through the points $(0, 0)$ and $(1, 1)$. For values of $x$ that are greater than 1, the higher the power, the higher the function value. For values of $x$ that are less than 1, the opposite trend holds: the smaller powers have higher function values. Notice also that the power functions with $p > 1$ are concave up, and the power functions with $0 < p < 1$ are concave down.

We have looked at several groupings of power functions. There are several trends that we observed, and to help you keep things straight I have summarized some of our observations in the table on page 129.

## Lesson 6-4: Transformations of Power Functions

In general, a power function is a function of the form $f(x) = ax^p$. When comparing two power functions, it is important to consider the asymptotic behavior of the power functions as well as the behavior of the functions for moderate values of $x$.

| $f(x) - x^p$ | Concavity on $(0, \infty)$ | Pattern Observed |
|---|---|---|
| $p > 1$ | Up | $(0, 1)$: Lower powers dominate |
| | | $(1, \infty)$: Higher powers dominate |
| $0 < p < 1$ | Down | $(0, 1)$: Lower powers dominate |
| | | $(1, \infty)$: Higher powers dominate |
| $p < 0$ | Up | $(0, 1)$: Values of $p$ with a higher *magnitude* dominate |
| | | $(1, \infty)$: Values of $p$ with a lower *magnitude* dominate |

For example, let's compare the functions $f(x) = 100x^2$ and $g(x) = x^4$. Both of these functions pass through the origin, and both of these functions are symmetric with respect to the $y$-axis.

We will first look at the asymptotic behavior of the two functions. For large values of $x$, $x^4 > 100x^2$. How large does $x$ have to be in order for this inequality to hold? Well, if $x^4 > 100x^2$, then $x^4 - 100x^2 > 0$. This is a polynomial inequality, which we learned how to analyze in Chapter 4. We can factor the polynomial: $x^4 - 100x^2 = x^2(x - 10)(x + 10)$, and then analyze the signs of this function. The sign chart for the function $f(x) = x^4 - 100x^2$ is shown in Figure 6.6. From the sign chart, we see that the inequality is true if $|x| > 10$.

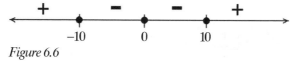

*Figure 6.6*

What this tells us is that for values of $x$ greater than 10 (or values of $x$ less than $-10$), $g(x) > f(x)$. Also, from the sign chart we see that if $|x| < 10$, $g(x) < f(x)$. The graphs of $f(x) = 100x^2$ and $g(x) = x^4$ are shown in Figure 6.7 on page 130. The effect of the constant 100 is that the point where the two power functions intersect has moved to the right. The functions $x^2$ and $x^4$ intersect at the origin and at $(1, 1)$, whereas the functions $100x^2$ and $x^4$ intersect at the origin and at $(10, 10^4)$.

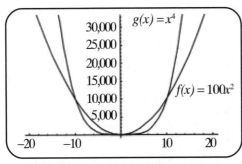

*Figure 6.7*

One of the reasons we study power functions is that, as we learned in the previous chapter, when $x$ is large enough, *every* polynomial or rational function will behave like a power function. Polynomials and rational functions *asymptotically* approach power functions. Understanding the behavior of power functions enables us to understand the behavior of polynomials and rational functions for extreme values of $x$.

Power functions can be used to calculate the area of a circle and the volume of a sphere. The area of a circle is given by the formula $A = \pi r^2$, and the volume of a sphere is given by the formula $V = \frac{4}{3}\pi r^3$. Power functions also appear frequently in physics and chemistry. In chemistry, Boyle's Law describes the relationship between the pressure of a gas, $P$, and its volume, $V$, and is given by the formula $P = kV^{-1}$. In physics, the distance, $d$, that an object falls during free fall can be modeled using the equation $d = \frac{1}{2}t^2$, where $t$ is the time that the object is falling. These power functions are often referred to as *functions of variation*.

## Example 1

Calculate the area of a circle with radius 5 in.

**Solution:** Using the formula $A = \pi r^2$, the area of the circle is

$A = \pi (5)^2 = 25\pi$ in$^2$.

## Example 2

Calculate the volume of a sphere of radius 3 ft.

**Solution:** Using the formula $V = \frac{4}{3}\pi r^3$, the volume of the

sphere is $V = \frac{4}{3}\pi (3)^3 = \frac{4}{3}\pi (27) = 36\pi$ ft$^3$.

# Exponential and Logarithmic Functions

In the last chapter, we learned about power functions. Power functions involve an independent variable raised to a power. Power functions, such as the function $f(x) = x^2$, are exponential expressions in which the base is the independent variable and the exponent is a constant. When we change things around so that the base is a constant and the independent variable is in the exponent, as with the function $g(x) = 2^x$, we create an exponential function. In this chapter we will examine and transform exponential functions.

## Lesson 7-1: Exponential Functions

An **exponential function** is a function of the form $f(x) = b^x$, where the base, $b$, is a positive constant. We will begin our examination of exponential functions by finding the y-intercept of an exponential function. Remember that any non-zero number raised to the power 0 is 1, so every exponential function passes through the point (0, 1).

If the base of an exponential function is greater than 1, the function will increase exponentially, or the function exhibits **exponential growth**. If the base of an exponential function is between 0 and 1, we say that the function decreases exponentially, or the function exhibits **exponential decay**. The value of the base will affect the growth rate of the exponential function. For exponential growth, a larger base will result in a larger growth rate. The closer the base of an exponential function is to 1, the less dramatic the exponential growth or decay of the function will be.

Through time, exponential growth is much more rapid than the growth of any power function. Because polynomials and rational functions behave as power functions do for large values of $x$, exponential growth will eventually overtake any polynomial, rational, or power function. Let's compare the growth of the functions $f(x) = x^2$ and $g(x) = 2^x$ for some positive values of $x$. I have evaluated these functions for several values of $x$ in the following table, and graphed these two functions in Figure 7.1.

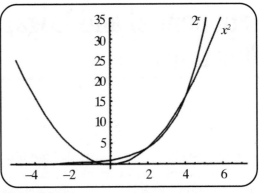

| $x$ | $f(x) = x^2$ | $g(x) = 2^x$ |
|---|---|---|
| 0 | 0 | 1 |
| 1 | 1 | 2 |
| 2 | 4 | 4 |
| 3 | 9 | 8 |
| 4 | 16 | 16 |
| 5 | 25 | 32 |
| 3 | 36 | 64 |
| 7 | 49 | 128 |

Figure 7.1

Notice that for small values of $x$, specifically for $0 \le x \le 4$, there are values of $x$ for which the exponential function is higher than the power function, and there are values of $x$ for which the power function is higher than the exponential function. But once $x > 4$, there will be no turning back. From that point on, $2^x$ will dominate $x^2$. Whenever you compare exponential growth and the growth of a power function, there will be a point beyond which the exponential function will remain higher than the power function.

An exponential function $f(x) = a^x$ will have a one-sided horizontal asymptote $y = 0$. If $a > 1$, then as $x$ approaches $-\infty$, $a^x$ gets closer and closer to the line $y = 0$. As $x$ approaches $+\infty$, $f(x) = a^x$ exhibits exponential growth, and the function becomes arbitrarily large. That's why the asymptote is one-sided. The function $f(x) = a^x$ will never cross the $x$-axis, however.

If $a < 1$, then $a^x$ represents exponential decay, and as $x$ approaches $+\infty$, $a^x$ will get closer and closer to the line $y = 0$. As $x$ approaches $-\infty$, an exponential decay function will become arbitrarily large.

An exponential function can be transformed by translation, reflection, expansion, and contraction. Starting with the function $f(x) = 2^x$, we can shift the function vertically (up or down) by $c$ units by adding $c$ to the function: the graph of $2^x + c$ is the graph of $2^x$ moved up or down $c$ units, depending on the sign of $c$. When an exponential function is shifted up or down $c$ units, its horizontal asymptote is also shifted up or down $c$ units. The function $2^x + c$ has a horizontal asymptote $y = c$. We can shift the function horizontally (left or right) by $c$ units by adding $c$ to the argument of the function. The graph of $2^{x+c}$ is the graph of $2^x$ moved to the left or right, depending on the sign of $c$, $c$ units. When an exponential function is shifted left or right, the horizontal asymptote doesn't change. If you move a horizontal line to the left or the right, you essentially have the same horizontal line. Notice that using our rules for exponents, we can simplify $2^{x+c}$: $2^{x+c} = 2^x 2^c$. Because $2^c$ is just a constant, this has the same effect as stretching or contracting, depending on whether $2^c > 1$ or $2^c < 1$, the graph of $2^x$.

We can also reflect the graph of an exponential function with respect to the $x$-axis and with respect to the $y$-axis. To reflect the graph of $2^x$ across the $x$-axis, multiply the function by $-1$. The graph of $-2^x$ is the graph of $2^x$ reflected across the $x$-axis. Use the order of operations carefully when you interpret $-2^x$. The base of the exponent is 2, and to evaluate this function you first raise 2 to the power $x$ and then multiply by $-1$. We can reflect the graph of $2^x$ across the $y$-axis by multiplying the independent variable by $-1$. The graph of $2^{-x}$ is the graph of $2^x$ reflected about the $y$-axis. We can simplify $2^{-x}$ using our rules for exponents:

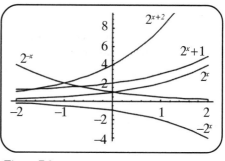

$2^{-x} = \left(2^{-1}\right)^x = \left(\frac{1}{2}\right)^x$. Reflecting the function $f(x) = 2^x$ across the $y$-axis turns our exponential growth into exponential decay. Reflecting an exponential function across the $y$-axis will not change the horizontal asymptote of the function. If you reflect an exponential function

*Figure 7.2*

across the $x$-axis, you must also reflect its horizontal asymptote. Figure 7.2 shows graphs of these transformations.

Exponential functions can also be used to model certain situations, but we need to make a slight modification. Exponential functions of the form $f(x) = b^x$ must pass through the point $(0, 1)$, which is too restrictive. To loosen things up, we will use a function of the form $f(x) = a \cdot b^x$. Notice that the $y$-intercept of the function $f(x) = a \cdot b^x$ is $(0, a)$. Fitting an exponential model to two data points involves using the two points to find the constants $a$ and $b$.

## Example 1

A cell culture has a bacteria population of 50 cells. One hour later, the cell population is 400. Find an exponential model for the cell population as a function of time, and use the model to predict the population in 8 hours.

**Solution:** We will use an exponential model of the form $P(t) = a \cdot b^t$ for this situation, where $P$ represents the population and $t$ represents the elapsed time, in hours. We have two data points: (0, 50) and (1, 400). We can substitute each data point into our model to generate two equations that only involve $a$ and $b$, and solve for the constants. Using the data points, we have:

$50 = a \cdot b^0$ and $400 = a \cdot b^1$.

The first equation can be used to solve for $a$ immediately: $50 = a$. Now that we know the value of $a$, we can use it in the second equation and solve for $b$:

$400 = a \cdot b^1$

$400 = 50 \cdot b$

$b = 8$

The exponential model is $P(t) = 50 \cdot 8^t$. We can use this model to predict the population when $t = 8$:

$P(t) = 50 \cdot 8^t$

$P(8) = 50 \cdot 8^8$

$P = 838,860,800$

In 8 hours, the population is 838,860,800 cells!

---

Exponential models are also useful when calculating compound interest. If an amount of money $P$, called the principal, is invested at an interest rate $i$ per time period, then after one time period, the interest is $P \cdot i$, and the amount of money, $A$ (including principal *and* interest) are

$$A = P + P \cdot i = P(1 + i).$$

If the principal and interest are reinvested for another time period, the amount of money will now be:

$$A = P(1 + i) + P(1 + i) \cdot i.$$

The first term in the equation is the amount reinvested, and the second term in the equation is the interest earned during the second time period. We can simplify this equation further:

$$A = P(1 + i) + P(1 + r) \cdot i$$
$$A = P(1 + i)^2 .$$

In general, after $n$ time periods, the amount of money will be

$$A = P(1 + i)^n.$$

Usually, interest rates quoted are *annual* interest rates, and interest is compounded a certain number of times per year. If the annual interest rate is $r$ and interest is compounded $n$ times per year, then the interest rate in each time period is $i = \frac{r}{n}$ and there are $n \cdot t$ time periods in $t$ years. Using this information, we obtain a general formula to calculate the amount of money after $t$ years:

$$A = P\left(1 + \frac{r}{n}\right)^{n \cdot t},$$

where $A$ is the amount of money after $t$ years, $P$ is the principal amount invested, $r$ is the annual interest rate (as a decimal), $n$ is the number of times that interest is compounded per year, and $t$ is the number of years that the money is invested. We can use this formula to calculate $A$, $P$, or $r$. We will be able to solve for $t$ in Lesson 7-5.

### Example 2

If $1,000 is invested at an annual rate of 10% per year compounded monthly, find the amount in the account after 5 years.

**Solution:** We are given that $P = 1,000$, $r = 0.10$, $n = 12$, and $t = 5$. Substituting for these variables, we can find $A$:

$$A = P\left(1 + \frac{r}{n}\right)^{n \cdot t}$$

$$A = 1,000\left(1 + \frac{0.1}{12}\right)^{12 \cdot 5}$$

$$A = 1,645.31$$

The account will have $1,645.31 after 5 years.

Homework Helpers: Pre-Calculus

136

## Example 3

Find the amount you would have to deposit into an account earning 8% annual interest compounded quarterly if you want the account to be worth $5,000 at the end of 10 years.

**Solution:** We are given that $A = 5,000$, $r = 0.08$, $n = 4$, and $t = 10$. Substituting in for these variables, we can find $P$:

$$A = P\left(1 + \frac{r}{n}\right)^{n \cdot t}$$

$$5,000 = P\left(1 + \frac{0.08}{4}\right)^{4 \cdot 10}$$

$$5,000 = P(2.208)$$

$$P = 2264.45$$

An initial amount of $2,264.45 would have to be invested.

---

The interest earned will increase as the number of times that interest is compounded increases. If we take things to an extreme and imagine that interest is being compounded continuously, the formula used to calculate $A$ will change. In this situation, we are looking at the asymptotic behavior of $A$ as $n$, the number of times we compound interest approaches infinity. If we let $m = \frac{n}{r}$, then $\frac{r}{n} = \frac{1}{m}$ and $n \cdot t = m \cdot r \cdot t$. We can substitute these expressions into our formula and simplify:

$$A = P\left(1 + \frac{r}{n}\right)^{n \cdot t}$$

$$A = P\left(1 + \frac{1}{m}\right)^{m \cdot r \cdot t}$$

$$A = P\left[\left(1 + \frac{1}{m}\right)^{m}\right]^{r \cdot t}$$

Now, as the number of times the interest is compounded heads towards infinity, so does $m$. And as $m \to \infty$, the expression $\left(1 + \frac{1}{m}\right)^{m}$ does something interesting. Notice that in the expression $\left(1 + \frac{1}{m}\right)^{m}$, both the base and the exponent are changing. The exponent $m$ will try to make the expression exhibit exponential growth (because the base is greater than 1).

But as $m$ increases, the base gets closer to 1. And the closer to 1 the base becomes, the slower the growth of the function. These two actions are contradictory, and instead of the function growing exponentially, it will actually taper off and asymptotically approach a fixed constant. This fixed constant is an irrational number that appears in some surprising places. It is given the name (or letter) $e$. This isn't the first time that mathematicians have referred to special numbers by a letter: pi, $\pi$, is a letter of the Greek alphabet used to represent the ratio of the circumference to the diameter of any circle. We can describe $e$ using the ideas of compound interest. If you deposit $1 in an account that earns interest at an annual rate of 100% compounded continuously, the amount of money in the account after 1 year will be $e$. The numerical value of $e$ is 2.71828.... It will never repeat in a pattern, as it is an irrational number.

To continue with our original investigation, we were analyzing the function

$$A = P\left[\left(1+\tfrac{1}{m}\right)^m\right]^{r \cdot t} \text{ as } m \to \infty.$$

We can use the fact that $\left(1+\tfrac{1}{m}\right)^m \to e$ as $m \to \infty$ to obtain the formula:

$$A = Pe^{rt}$$

### Example 4

If $1,000 is invested at an annual rate of 10% per year compounded continuously, find the amount in the account after 5 years.

**Solution:** Use the equation $A = Pe^{rt}$ to find $A$ when $P = 1,000$, $r = 0.10$, and $t = 5$:

$A = Pe^{rt}$

$A = 1,000e^{0.10 \cdot 5}$

$A = 1,648.72.$

The account will have $1,648,72 after 5 years.

Notice that this answer is a little larger than the answer in Example 2; realize that compounding continuously is as good as it gets. It should come as no surprise that financial institutions will compound often on loans and credit card balances, and compound as infrequently as possible on savings and checking accounts.

You may have noticed that when banks post interest rates, they give two values: an interest rate and a rate called the Annual Percentage Yield, or APY. When comparing two loans, or two investments, it is important to consider the interest rate and the APY. Posting one rate without the other could be misleading. The APY that corresponds to a given rate $r$ compounded $n$ times per year is given by $APY = \left(1 + \frac{r}{n}\right)^n$, and if interest is compounded continuously, the APY is calculated using the formula $APY = e^r$

## Example 5

Which offers a higher APY, 10% compounded quarterly or 9.9% compounded daily?

**Solution:** The APY for the first option is

$$APY = \left(1 + \frac{.1}{4}\right)^4 = 1.10381,$$

and the APY for the second option is

$$APY = \left(1 + \frac{.099}{365}\right)^{365} = 1.10405.$$

The lower interest rate with interest compounded daily is the better option in this case.

---

Once you have created several exponential models, you should begin to notice a pattern. For example, an exponential model for the amount of a radioactive substance present after $t$ days if initially there are 50 grams and the half-life of the substance is 10 days, has the form $A = 50\left(\frac{1}{2}\right)^{t/10}$. If the a bacteria sample has 300 bacteria present initially and the population doubles every 120 minutes, then an exponential model for the bacteria population present after $t$ minutes has the form $A = 300(2)^{t/120}$. Notice what is similar about these two situations. In the first example, we were given the half-life of the substance, and the base of the corresponding exponential model was ½. In the second example, we were given the doubling time of the population, and the base of the corresponding exponential model was 2. The constant in front of the exponent reflects the initial amount of the substance or of the population. And the exponent involves $t$ divided by the half-life or the doubling time. Following this pattern can help you generate some exponential models very quickly. If you forget about this pattern, you can always determine two points and fit an exponential model of the form $f(x) = a \cdot b^t$. It may look slightly different than the one that we generate using this shortcut, but the two equations will be equivalent. If you work out the exponential models both ways, you'll see how the two forms relate to each other.

## Lesson 7-1 Review

1. The half-life of a radioactive material is the time that it takes for half of a sample to undergo radioactive decay. If a radioactive material has a half-life of 30 minutes, construct an exponential model to determine the amount of a substance as a function of time, assuming that there are initially 20 grams of the material present. Use this model to predict the amount (in grams) of this radioactive material present in 45 minutes.

2. If $25,000 is invested at an annual rate of 5% per year compounded quarterly, find the amount in the account after 2 years.

3. Find the amount that you would have to deposit into an account earning 12% annual interest compounded continuously if you want the account to be worth $3,000 at the end of 6 years.

# Lesson 7-2: Logistic Growth

The problem with using an exponential function to model populations is that, according to the model, an exponential increase in population means that the population can grow arbitrarily large. In reality, population growth is limited by a variety of factors. Available resources may have a significant effect on a population. Scarcity of food, for example, may result in a decrease in the growth rate, or a decrease in the population if it is severe enough. If there is not enough space to support the population, the population growth rate may decrease. A better model for population dynamics is logistic growth.

A logistic growth function is a function of the form : $L(t) = \dfrac{K}{1 + ae^{-bt}}$ where $K$, $a$, and $b$ are constants. We can interpret $L$ as the population as a function of time, $t$. This function has the property that when $t$ is small, the exponential expression in the denominator is much larger than 1, so $1+ae^{-bt} \sim ae^{-bt}$, and overall, $L(t) \sim \frac{K}{ae^{-bt}} = \frac{K}{a}e^{bt}$. In other words, initially the resources are sufficient to support population growth. As time increases, however, the resource limitations begin to affect the population growth, and as $t$ becomes large, we have:

$1+ae^{-bt} \sim 1$, and overall, $L(t) \sim \frac{K}{1} = K$.

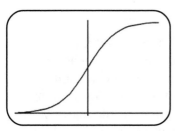

*Figure 7.3*

Now the population is leveling off, and reaching a maximum sustainable population $K$. The constant $K$ is called the carrying capacity of the environment. The graph of a logistic growth curve is shown in Figure 7.3.

The logistic growth function illustrates how we can transform an exponential function to more accurately model an observable phenomenon. Logistic growth models are accurate in modeling the population of a species when there are limiting resources, such as food or space, that prevent the population from increasing without bound.

## Lesson 7-3: Hyperbolic Functions

Another popular transformation of exponential functions involves adding and subtracting them. The hyperbolic functions play an important role in physics and engineering. The **hyperbolic sine** function, abbreviated sinh, is defined as:

$$\sinh(x) = \frac{e^x - e^{-x}}{2}.$$

The **hyperbolic cosine** function, abbreviated cosh, is defined as:

$$\cosh(x) = \frac{e^x + e^{-x}}{2}.$$

We can analyze the formulas for these functions to understand them better. Notice first that $\sinh(0) = \frac{e^0 - e^0}{2} = 0$ and $\cosh(0) = \frac{e^0 + e^0}{2} = 1$. The hyperbolic sine function passes through the origin, while the $y$-intercept of the hyperbolic cosine is (0, 1). The formulas for both of these functions are symmetric, and it should not surprise you to learn that the graphs of these functions will also exhibit symmetry. Notice that

$$\sinh(-x) = \frac{e^{-x} - e^{-(-x)}}{2} = \frac{e^{-x} - e^{x}}{2} = -\left(\frac{e^{x} - e^{-x}}{2}\right) = -\sinh(x).$$

In other words, the hyperbolic sine function is an odd function, and is symmetric about the origin. We can also analyze the hyperbolic cosine function and discover its symmetry:

$$\cosh(-x) = \frac{e^{-x} + e^{-(-x)}}{2} = \frac{e^{-x} + e^{x}}{2} = \frac{e^{x} + e^{-x}}{2} = \cosh(x).$$

We see that the hyperbolic cosine function is an even function, and is symmetric about the *y*-axis.

In addition to the symmetry of the hyperbolic functions, it is worthwhile to examine the asymptotic behavior of these functions. As *x* gets large, $e^x$ gets large and $e^{-x}$ becomes negligible. As $x \to \infty$, both sinh(*x*) and cosh(*x*) behave like $e^x$. We can write that as $x \to \infty$, sinh(*x*) ~ $e^x$ and cosh(*x*) ~ $e^x$. If *x* is large in magnitude but is negative, then $e^x$ becomes negligible and $e^{-x}$ is large and dominates. As $x \to -\infty$, sinh(*x*) behaves like $-e^{-x}$ and cosh(*x*) behaves like $e^{-x}$. We can write that as $x \to -\infty$, sinh(*x*) ~ $-e^{-x}$ and cosh(*x*) ~ $e^{-x}$. Graphs of the hyperbolic sine and cosine curves are shown in Figure 7.4.

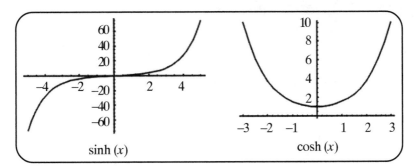

sinh (*x*)                cosh (*x*)

*Figure 7.4*

There are several identities that the hyperbolic functions satisfy. Examining one of these identities will help uncover why these functions are called *hyperbolic* functions. Let's compare $[\cosh(x)]^2$ and $[\sinh(x)]^2$ and make some observations:

$$[\cosh(x)]^2 = \left(\frac{e^x + e^{-x}}{2}\right)^2 = \frac{e^{2x} + 2 + e^{-2x}}{4}$$

$$[\sinh(x)]^2 = \left(\frac{e^x - e^{-x}}{2}\right)^2 = \frac{e^{2x} - 2 + e^{-2x}}{4}$$

Notice that $[\cosh(x)]^2$ and $[\sinh(x)]^2$ are very similar to each other. In fact, the only difference between them is the sign of the middle term. If we subtract one from the other, we have:

$$[\cosh(x)]^2 - [\sinh(x)]^2 = \frac{e^{2x} + 2 + e^{-2x}}{4} - \frac{e^{2x} - 2 + e^{-2x}}{4} = \frac{4}{4} = 1.$$

This second identity is the reason that these functions are called hyperbolic functions. The graph of the equation $\frac{x^2}{a^2} - \frac{y^2}{b^2} = 1$ is a hyperbola, and the second identity we established is $[\cosh(x)]^2 - [\sinh(x)]^2 = 1$, which is remarkably similar to the equation for a hyperbola. We will examine hyperbolas in more detail in Chapter 10.

The names of the hyperbolic functions may remind you of the trigonometric functions $\sin(x)$ and $\cos(x)$. There are many similarities between these two types of functions, which motivated how these functions were named.

## Lesson 7-4: Inverse Functions

A function is used to represent the relationship, or dependence, between one quantity and another. The rate at which a cricket chirps is related to the temperature of its environment. Under certain circumstances, we can use a cricket as a thermometer. In this situation, we are looking at the environmental temperature as a function of the chirp rate. Looking at things from the reverse perspective can be thought of mathematically as inverting a function.

There are rules for how to invert a function, and there are conditions under which a function cannot be inverted. We denote the inverse of the function $f$ by $f^{-1}$. When dealing with rational and exponential expressions, we use the superscript $-1$ to denote the reciprocal of a number. Now we are using it to denote the *inverse* of a function. There is a reason that we are reusing this notation. The *reciprocal* of a number $a$ is the number to *multiply a* by to get 1 (the *multiplicative identity*). The **inverse** of a function $f$ is a function that you *compose* with $f$ to get the *identity function*.

An identity is something that doesn't change the input. The *additive* identity is the number that doesn't change the input under addition: the additive identity is 0. The *multiplicative* identity is the number that doesn't change the input under multiplication: the multiplicative identity is 1. The identity *function* is the function that does not change the input of the function: the identity function is $f(x) = x$. Notice that the input of the function is the same as the output of the function. The inverse of a function $f$ is the function, denoted $f^{-1}$, that has the following relationship to $f$:

$$(f^{-1} \circ f)(x) = x \text{ and } (f \circ f^{-1})(x) = x.$$

If a function has an inverse, the function is **invertible**.

As I mentioned earlier, the inverse of a function involves a change in perspective: the output becomes the input and the input becomes the output. If $f$ is an invertible function with domain $X$ and range $Y$, then $f^{-1}$ will be a function with domain $Y$ and range $X$. When inverting a function, the role of the independent variable, $x$, and the role of the dependent variable, $y$ or $f$, switch. This concept will be useful in actually finding the inverse of an invertible function. If an invertible function $f$ passes through the point $(a, b)$, then its inverse, $f^{-1}$, will pass through the point $(b, a)$: the input becomes the output and the output becomes the input. If $f$ represents the chirp rate as a function of temperature, where temperature is measured in degrees Fahrenheit, and $f(60) = 100$, then this means that at a temperature of 60° F, the chirp rate is 100 chirps per minute. From this, we know that $f^{-1}(100) = 60$, meaning that a chirp rate of 100 chirps per minute corresponds to an environmental temperature of 60° F. If a function $f$ is invertible, and $f(a) = b$, then $f^{-1}(b) = a$. In other words, if $f$ is invertible, then $f^{-1}(r) = s$ means the same thing as $f(s) = r$.

Not all functions have inverses. Suppose that $f(2) = 4$ and $f(-2) = 4$. Which value would you choose for $f^{-1}(4)$? Some people may decide that $f^{-1}(4) = 2$ and other people may prefer that $f^{-1}(4) = -2$. Because $f^{-1}(4)$ is not unique, we avoid problems by declaring that $f$ is not invertible. Our understanding of symmetry allows us to conclude that every even function is not invertible. Remember that an even function satisfies the relationship $f(-x) = f(x)$, so if $f(a) = b$, then $f(-a) = b$ and there is no unique choice for $f^{-1}(b)$.

A function will be invertible if it is one-to-one. A function is **one-to-one** if no two elements in the domain have the same image. In other words, if $a$ and $b$ are in the domain of $f$, and $a \neq b$, then $f(a) \neq f(b)$. Another way to look at this situation is that if $f(a)$ and $f(b)$ are two points in the range with $f(a) = f(b)$, then $a = b$. Determining whether a function has an inverse is equivalent to determining whether the function is one-to-one.

## Example 1

Is the function $f(x) = x^2 + 4$ invertible?

**Solution:** If a function is invertible, it will be one-to-one. Use the definition to determine whether $f(x) = x^2 + 4$ is one-to-one. Suppose that $f(a)$ and $f(b)$ are two points in the range that are equal: $f(a) = f(b)$.

Because $f(x) = x^2 + 4$, $f(a) = a^2 + 4$ and $f(b) = b^2 + 4$.

Now, $f(a) = f(b)$, so we have $a^2 + 4 = b^2 + 4$.

We can simplify this equation and discover the relationship between $a$ and $b$:

$a^2 + 4 = b^2 + 4$

$a^2 = b^2$

$a^2 - b^2 = 0$

$(a - b)(a + b) = 0$

$a = b$ or $a = -b$

We have a choice in how $a$ and $b$ are related: either $a = b$ or $a = -b$. This function is not one-to-one because of this choice. Another way to look at this function is to recognize that $f(x) = x^2 + 4$ is an even function, and even functions are not invertible.

## Example 2

Is the function $f(x) = 4x + 2$ invertible?

**Solution:** If a function is invertible, it will be one-to-one. Use the definition to determine whether $f(x) = 4x + 2$ is one-to-one.

Suppose that $f(a)$ and $f(b)$ are two points in the range that are equal: $f(a) = f(b)$.

Because $f(x) = 4x + 2$, $f(a) = 4a + 2$ and $f(b) = 4b + 2$.

Now, $f(a) = f(b)$, so we have $4a + 2 = 4b + 2$.

We can simplify this equation and discover the relationship between $a$ and $b$:

$4a + 2 = 4b + 2$

$4a = 4b$

$a = b$

In this case, we do not have a choice in how $a$ and $b$ are related: $a = b$. This function is one-to-one.

---

If a function is not invertible, we can restrict the domain so that it will be invertible. For example, the function $f(x) = x^2 + 4$ is not one-to-one, so it is not invertible. However, if we restrict the domain of $f$ to be the non-negative real numbers, then $f$ will be invertible. With this restriction, power functions will have inverses: the inverse of $f(x) = x^p$ will be $f^{-1}(x) = x^{1/p}$. The inverse of a power function is a root. We use this relationship without much thought in algebra. In order to solve the equation $x^2 = 4$, most people simply "take the square root of both sides and throw in a $\pm$." This process actually involves restricting the domain of the function $f(x) = x^2$ to the non-negative real numbers, and using its inverse function $f^{-1}(x) = \sqrt{x}$ to find one value of $x$ that satisfies the equation. Then the other solution to the equation is found from symmetry; $f(x) = x^2$ is an even function, so the other solution will be on the other side of the $y$-axis, hence the $\pm$. There is no need to change *how* you solve a quadratic equation, but it's always nice to be aware of the reason *why* you get the correct answer when you take shortcuts.

If a function is defined by a formula and is invertible, it is sometimes possible to find a formula for the inverse function. The key to finding a formula for the inverse function is to realize that the role of $x$ and the role of $y$ switch when you invert a function. The domain becomes the range, and the range becomes the domain. To find the inverse of a function $y = f(x)$, first switch $x$ and $y$, and then solve for $y$. To switch $x$ and $y$, everywhere you see an $x$, write a $y$, and everywhere you see a $y$, replace it with an $x$.

## Example 3

Find the inverse of $f = \{(1, 2), (3, 4), (5, 6)\}$.

**Solution:** To find the inverse, switch the domain and the range:
$f^{-1} = \{(2, 1), (4, 3), (6, 5)\}$.

---

## Example 4

Find the inverse of the function $f(x) = 2x + 1$.

**Solution:** Rewrite the function as $y = 2x + 1$. Switch $x$ and $y$ and then solve for $y$: $y = 2x + 1$ becomes $x = 2y + 1$.

Now we can solve for $y$:

$$x = 2y + 1$$

$$2y = x - 1$$

$$y = \frac{x-1}{2}$$

Thus $f^{-1}(x) = \frac{1}{2}(x-1)$.

Notice that the function $f(x) = 2x + 1$ takes the input, doubles it, and then adds 1. The inverse function, $f^{-1}$, will just undo what the function $f$ does, but in reverse order. First, $f^{-1}$ undoes the addition of 1 by subtracting 1. Then $f^{-1}$ undoes the multiplication by 2, by dividing by 2. You can see this in the formula for $f^{-1}$: $f^{-1}(x) = \frac{1}{2}(x-1)$.

## Example 5

Find the inverse of the function $f(x) = \frac{1}{x+2}$.

**Solution:** Rewrite the function as $y = \frac{1}{x+2}$.

Switch $x$ and $y$ and then solve for $y$: $y = \frac{1}{x+2}$ becomes $x = \frac{1}{y+2}$.

Now we can solve for $y$ by taking the reciprocal of both sides and then isolating $y$:

$$x = \frac{1}{y+2}$$

$$\frac{1}{x} = y + 2$$

$$y = \frac{1}{x} - 2$$

Thus $f^{-1}(x) = \dfrac{1}{x} - 2$.

_____

Again, if we think about what the function $f$ does, we can reason out what $f^{-1}$ will do. The function $f$ takes the input and first adds 2, then takes the reciprocal of the result. The function $f^{-1}$ will undo what $f$ does, in reverse order: it will first take the reciprocal of the input, and then subtract 2. If you can break apart the effects of $f$, then you can find $f^{-1}$ by undoing what $f$ does, in reverse order.

As I mentioned earlier, not all functions have inverses. But for many functions, restricting the domain will enable us to define an inverse.

The graph of an invertible function and its inverse are closely related to each other. The graph of the inverse of a function is a reflection of the function across the line $y = x$. Figure 7.5 shows the graph of a function and its inverse on the same axes. Graphing the inverse of a function takes some practice. Keep in mind that the line $y = x$ acts like a mirror; any points in Quadrant II are moved to Quadrant IV, points in Quadrant I stay in Quadrant I, and points in Quadrant III stay in Quadrant III.

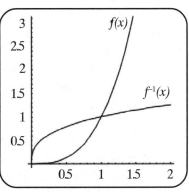

Figure 7.5

## Lesson 7-4 Review

1. Are the following functions invertible?
   a. $f(x) = \cosh(x)$
   b. $g(x) = x^2$
   c. $h(x) = 3x + 1$

2. Find the inverse of the following functions:
   a. $f(x) = 3x - 4$

   b. $g(x) = \dfrac{2}{x - 3}$

3. If $f = \{(1, 3), (-2, 6), (3, 5)\}$, find the domain of $f^{-1}$.

# Lesson 7-5: Logarithmic Functions

In Lesson 7-1, we learned how to create a function to model the amount of money in an account as a function of time, given that the initial deposit is $1,000 and interest is compounded quarterly at an annual rate of 10%:

$$A(t) = 1,000(1.025)^{4t}.$$

Now, suppose we wanted to determine when the balance in the account reaches $1,500. In other words, we want to find the value of $t$ for which

$$1,500 = A(t) = 1,000(1.025)^{4t}.$$

Because this exponential function is always increasing, and it starts out with an initial balance of $1,000, there will be one unique value $t$ for which $A(t) = 1,500$. One way to find this value of $t$ is through trial and error. Substitute various values of $t$ and adjust your focus accordingly. For a starting point, the amount in the account will reach $1,500 after $500 in interest is earned. Because 10% of $1,000 is $100, we expect that it should take no more than 5 years to earn $500 in interest, to bring the balance up to $1500. Start by evaluating $A(t)$ for $t = 3$, $t = 4$, and $t = 5$ as shown in the chart here.

| $t$ | $A(t)$ |
|-----|--------|
| 3 | 1,344.89 |
| 4 | 1,484.51 |
| 5 | 1,638.62 |

The time that it will take is somewhere between 4 and 5 years. To obtain a more accurate solution, evaluate $A(4.5)$: $A(4.5) = 1,559.66$. Next, evaluate $A(4.25)$, and continue to hone in on a more accurate estimate for $t$. This method of trial and error will always work, but it may be time consuming. Solving the equation $A(t) = 1,500$ is equivalent to evaluating $A^{-1}(1,500)$. There is a systematic approach to solving these problems: find a formula for $A^{-1}$. The logarithmic function is the inverse of the exponential function.

The exponential function $f(x) = b^x$ is an invertible function, and we will find its inverse by first switching $x$ and $y$ and then solving for $y$:

$$y = b^x$$
$$x = b^y$$

At this point, we are stuck. We don't know how to isolate the exponent from the base. We define the **logarithmic function** to be the inverse of the exponential function, and write:

$$y = \log_b x \quad \leftrightarrow \quad x = b^y$$

In the *exponential* function $f(x) = b^x$, $b$ is the base; in the logarithmic function $y = \log_b x$, $b$ is the base of the *logarithm*. These functions are inverses of each other, and $y = \log_b x$ *means the same thing as* $b^y = x$. In words, $x = b^y$ means that $y$ is the power you raise $b$ to in order to get $x$. Because $y = \log_b x$ means the same thing as $x = b^y$, the equation $y = \log_b x$ *also* means that $y$ is the power you raise $b$ to in order to get $x$.

For example, to evaluate $x = \log_2 2$, $x$ is the power you raise 2 to in order to get 2. Well, $2^1 = 2$, because 1 is the power you raise 2 to in order to get 2: $\log_2 2 = 1$. To evaluate $x = \log_2 8$, think of the power that 2 must be raised to in order to get 8: $2^3 = 8$, so 3 is the power that 2 must be raised to in order to get 8: $\log_2 8 = 3$.

To understand the properties of the logarithmic function, it is best to start with the exponential function. The domain of the exponential function is all real numbers, and the range is the set of positive real numbers. When you invert a function, the domain becomes the range and the range becomes the domain. Because the logarithmic function is the inverse of the exponential function, the domain of the logarithmic function is the set of positive real numbers, and the range is the set of all real numbers. Remember that every exponential function $f(x) = a^x$ has a $y$-intercept of $(0, 1)$. From this, we know that every logarithmic function $f(x) = \log_b x$ has an $x$-intercept of $(1, 0)$. Also, because every exponential function $f(x) = a^x$ has a horizontal asymptote $y = 0$, so the logarithmic function has a vertical asymptote $x = 0$. The graph of $f(x) = \log_2 x$ is shown in Figure 7.6. Its graph is the reflection of the function $f(x) = 2^x$ reflected across the line $y = x$.

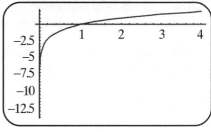

Some important rules for simplifying logarithms are summarized as follows. These rules stem from the rules for simplifying exponents. These properties are valid only when $A$ and $B$ are in the domain of the logarithmic function. In other words,

*Figure 7.6*

these properties are only true if $A$ and $B$ are positive real numbers:

▸ $\log_b (AB) = \log_b (A) + \log_b (B)$

▸ $\log_b \left( \dfrac{A}{B} \right) = \log_b (A) - \log_b (B)$

▶ $\log_b (A^p) = p \cdot \log_b (A)$

▶ $\log_b (1) = 0$

▶ $\log_b (b^p) = p$

▶ $b^{\log_b A} = A$

There are two important bases for a logarithm: base 10 (referred to as the **common logarithm**) and base $e$ (referred to as the **natural logarithm**). Technically, we can work with any base, but common logarithms are used in science, and the natural logarithm appears throughout science and mathematics. Numerical values for expressions that involve common and natural logarithms can be evaluated using a standard calculator. If the base of a logarithm is not 10 or $e$, then the base of a logarithm can be changed using the change of base formula:

$$\log_a x = \frac{\log_b x}{\log_b a}$$

If you need to evaluate the expression $\log_6 3$, it is a simple matter of evaluating $\frac{\log_{10} 3}{\log_{10} 6}$. You would get the same result if you evaluated $\frac{\log_e 3}{\log_e 6}$. The natural logarithmic function is written $\ln(x)$, and the common logarithm function is written $\log (x)$. If the base of the logarithm is other than 10 or $e$, then the base must be included in the logarithmic expression.

## Example 1

Find the domain, range, and asymptotes of $f(x) = \log_2 (x - 1)$.

**Solution:** We may not be familiar with the logarithmic function, but because it is the inverse of the exponential function (which we understand fairly well), if we find the inverse of $f(x) = \log_2 (x - 1)$ (which will be an exponential function), we will be able to answer the question. To find the inverse of $f(x) = \log_2 (x - 1)$, switch $x$ and $y$ and solve for $y$:

$$f(x) = \log_2 (x - 1)$$

Switch $x$ and $y$                  $x = \log_2 (y - 1)$

Write the equation in exponential form    $2^x = y - 1$

Add 1 to both sides                   $y = 2^x + 1$

The exponential function $y = 2^x + 1$ has domain $(-\infty, \infty)$, range $(1, \infty)$, and horizontal asymptote $y = 1$. From this we know that the domain of $f(x) = \log_2(x - 1)$ is $(1, \infty)$, the range is $(-\infty, \infty)$, and the function has a vertical asymptote $x = 0$.

## Lesson 7-5 Review

1. Find the domain, range, and asymptotes of $f(x) = \log_2(2 - x)$.

# Lesson 7-6: Exponential and Logarithmic Equations

Solving exponential and logarithmic equations involve similar steps. First, try to isolate the exponential or logarithmic expressions on one side of the equation. If the equation involves solving for the exponent, take the logarithm of both sides (either the common logarithm or the natural logarithm) and use the properties of logarithms to simplify the new equation. If the equation involves solving for the argument of a logarithm function, rewrite the equation in exponential form and then simplify.

We will start by solving exponential equations. An **exponential equation** is an equation in which the variable occurs in the exponent.

## Example 1

Solve the equation $3^x = 8$ to six decimal places.

**Solution:** The exponential expression is already isolated, so we'll take the natural logarithm of both sides and then solve for $x$:

$$3^x = 8$$

Take the natural logarithm of both sides

$$\ln(3^x) = \ln(8)$$

Use the properties of logarithms to simplify

$$x \cdot \ln(3) = \ln(8)$$

Divide both sides by $\ln(3)$
$$x = \frac{\ln(8)}{\ln(3)} = \frac{2.07944}{1.09861} = 1.89279$$

## Example 2

Solve the equation $4^{2x+1} = 9$ to six decimal places.

**Solution:** The exponential expression is already isolated, so we'll take the common logarithm of both sides and then solve for $x$:

$$4^{2x+1} = 9$$

Take the common logarithm of both sides

$$\log(4^{2x+1}) = \log(9)$$

Use the properties of logarithms to simplify

$$(2x+1)\log(4) = \log(9)$$

Divide both sides by log (4)
$$(2x+1) = \frac{\log(9)}{\log(4)}$$

Subtract 1 from both sides
$$2x = \frac{\log(9)}{\log(4)} - 1$$

Divide both sides by 2
$$x = \frac{1}{2}\left(\frac{\log(9)}{\log(4)} - 1\right)$$

Use a calculator to evaluate the expression $x = 0.292481$

## Example 3

Solve the equation $x^2(2^x) - 6x(2^x) = 0$.

**Solution:** Both terms in this equation have an exponential component $2^x$ in common, so we can factor it out:

$x^2(2^x) - 6x(2^x) = (2^x)(x^2 - 6x)$.

Now we can solve the equation $(2^x)(x^2 - 6x) = 0$.

Because $2^x \neq 0$ for all $x$, the only way for this product to equal 0 is if the polynomial component is equal to 0: $(x^2 - 6x) = 0$. This equation can be solved by factoring:

$$x(x-6) = 0$$

$x = 0 \qquad x - 6 = 0$

$\qquad\qquad x = 6$

The solutions are $x = 0$ and $x = 6$.

## Example 4

Solve the equation $e^{2x} - 3e^x - 4 = 0$.

**Solution:** This problem doesn't seem to fit into the same category as the other problems, but with a simple substitution, it will. If we let $t = e^x$, then $t^2 = (e^x)^2 = e^{2x}$, and the equation to solve becomes $t^2 - 3t - 4 = 0$. This is a quadratic equation that can be solved by factoring:

$$t^2 - 3t - 4 = 0$$

$$(t - 4)(t + 1) = 0$$

$$t = 4 \qquad t = -1$$

Now that we know the values for $t$, we can solve for $x$ by using the substitution equation $t = e^x$ when $t = 4$:

$$t = e^x$$

Substitute $t = 4$ into the equation $\qquad e^x = 4$

Take the natural logarithm of both sides

$$\ln(e^x) = \ln(4)$$

Use the property of logarithms $\qquad x \cdot \ln(e) = \ln(4)$

$$\ln(e) = 1 \quad x = \ln(4)$$

Now we can examine what happens if $t = -1$:

$$t = e^x$$

$$e^x = -1$$

This equation has no solution, because $e^x > 0$ for all $x$.

The solution to the equation $e^{2x} - 3e^x - 4 = 0$ is $x = \ln(4)$.

---

We will now turn our attention to solving logarithmic equations. A **logarithmic** equation is an equation in which the variable occurs in the argument of a logarithmic function. To solve these equations, isolate the terms that involve logarithms on one side of the equation, and non-logarithmic terms on the other side. If there is more than one term that involves a logarithm, use the properties of logarithms to combine them. Then, write the equivalent equation in exponential form by using the relationship

$$y = \log_b x \quad \leftrightarrow \quad x = b^y.$$

Finally, solve the resulting equation. If you use the properties of logarithms to combine two or more logarithmic expressions, you *must* check your answers in the *original* problem. When you combine two or more logarithmic expressions, it is possible to introduce extraneous solutions that must later be rejected. An **extraneous solution** is a solution to an intermediate equation that is not a solution of the original equation. Extraneous solutions can be introduced when you combine two logarithmic expressions, when you square both sides of an equation, or whenever you perform an operation that is not invertible.

## Example 5

Solve the equation $\log_3 (x - 1) = 2$.

**Solution:** There is only one term that involves a logarithm, and it is already isolated, so we can write the equivalent equation in exponential form and solve it:

$\log_3 (x - 1) = 2 \qquad \leftrightarrow \qquad 3^2 = (x - 1)$

$9 = (x - 1)$

$x = 10$

## Example 6

Solve the equation $\ln(x + 1) = 1$.

**Solution:** There is only one term that involves a logarithm, and it is already isolated, so we can write the equivalent equation in exponential form and solve it:

$\ln(x + 1) = 1 \qquad \leftrightarrow \qquad e^1 = (x + 1)$

$e = (x + 1)$

$x = e - 1$

## Example 7

Solve the equation $\log_2 x + \log_2 (x + 2) = 3$.

**Solution:** Use the properties of logarithms to combine the two logarithmic expressions:

$\log_2 x + \log_2 (x + 2) = \log_2 [x(x + 2)]$.

Now, solve the equation $\log_2 [x(x+2)] = 3$ by writing the equivalent equation in exponential form:

$\log_2 [x(x+2)] = 3 \leftrightarrow 2^3 = [x(x+2)]$.

Now, solve the equation $2^3 = [x(x+2)]$:

| | |
|---|---|
| | $2^3 = [x(x+2)]$ |
| Expand both expressions | $8 = x^2 + 2x$ |
| Subtract 8 from both sides | $x^2 + 2x - 8 = 0$ |
| Factor | $(x+4)(x-2) = 0$ |
| Set each factor equal to 0 and solve | $x = -4 \text{ or } x = 2$ |

Because we used our properties of logarithms to combine two logarithmic expressions, we must make sure that the values of $x$ that satisfy the quadratic equation $2^3 = [x(x+2)]$ make sense in the *original* equation. If one or both of the values of $x$ do not make sense in the original equation, then we must discard those values of $x$, as they will be extraneous solutions. Substituting $x = -4$ into the original equation yields $\log_2 (-4) + \log_2 (-4+2)$, and these expressions do not make sense. Remember that the domain of the logarithm function is the set of positive real numbers. So $x = -4$ is an extraneous solution. Substituting $x = 2$ into the original equation yields $\log_2 (2) + \log_2 (2+2)$. Both of these expressions make sense, meaning that $x = 2$ is a valid solution to the original equation. Therefore, the only solution to the equation $\log_2 x + \log_2 (x+2) = 3$ is $x = 2$.

## Lesson 7-6 Review

Solve the following exponential and logarithmic equations:

1. $3^x = 4^{2x-1}$

2. $x^2(3^x) - 8x(3^x) + 12(3^x) = 0$

3. $e^{4x} - 5e^{2x} + 4 = 0$

4. $\log_4 (1-x) = 1$

5. $\ln(3x+2) = 2$

6. $\log_{20} (x+3) + \log_{20} (x+2) = 1$

7. $\log_3 (x+1) - \log_3 (x-1) = 1$

# Answer Key

## Lesson 7-1 Review

1. $A = 20\left(\frac{1}{2}\right)^{\frac{t}{30}}$; after 45 minutes, there are 7.07 grams.

2. $A = 25,000\left(1 + \frac{0.05}{4}\right)^{4 \cdot 2} = 2,7612.20$. The account has $27,612.20.

3. $3,000 = Pe^{0.12 \cdot 6} = 1,460.26$ You would have to invest $1,460.26.

## Lesson 7-4 Review

1. a. $f(x) = \cosh(x)$: no (it is an even function)
   b. $g(x) = x^2$: no (it is an even function)
   c. $h(x) = 3x + 1$: yes (it is a linear function)

2. a. $f^{-1}(x) = \frac{x+4}{3}$

   b. $g^{-1}(x) = \frac{2}{x} + 3$

3. The domain of $f^{-1}$ is $\{3, 5, 6\}$

## Lesson 7-5 Review

1. The domain of $f(x) = \log_2(2 - x)$ is $(-\infty, 2)$,
   the range is $(-\infty, \infty)$, and the vertical asymptote is $x = 2$.

## Lesson 7-6 Review

1. $x = 0.828144$

2. $x = 6$ or $x = 2$

3. $x = 0.693$ or $x = 0$

4. $x = -3$

5. $x = \frac{e^2 - 2}{3}$

6. $x = 2$

7. $x = 2$

# Systems of Equations and Inequalities

A **system of equations** refers to a *collection* of equations. A solution to a system of equations consists of all of the ordered pairs $(a, b)$ that satisfy all of the equations in the system. That means that any solution of a system of equations must satisfy all of the individual functions that make up the system. In other words, the solution of a system of equations consists of all points that lie at the intersection of all of the equations in the system.

A **system of linear equations** refers to a *collection* of linear equations. When solving systems of linear equations, it is important to keep in mind how many solutions the system has. The key to understanding how many answers you are looking for is to keep in mind the properties of lines. Recall that two points determine the equation of a line. If you have two lines, there are only a three possibilities:

▶ The two lines are parallel and do not intersect. In this case, there are no points that lie on both lines. In other words, there is no solution to the system of equations. The system of equations is **inconsistent**, meaning that there is no solution.

▶ The two lines are different representations of the same line, and the two equations are equivalent. If the two equations are equivalent then every point that satisfies one of the equations also satisfies the other equation. In other words, there are infinitely many solutions to the system of equations. The system of equations is **consistent**, because there is a solution. The system of equations is also considered to be **dependent**, because there is not a unique solution.

▶ The two lines are not parallel and the equations are not equivalent to each other. In this case, the two lines will intersect at one unique point and there is only one solution to the system of equations. The system of equations is consistent, because there is a solution. The system of equations is also considered to be **independent**, because the solution is unique.

We will see examples of each of these cases, though our focus will be on the third situation.

# Lesson 8-1: Solving Linear Systems Graphically

The most direct method for solving systems of equations is to graph all of the equations in that system and find the point of intersection. Because the solution of a linear system of equations must satisfy each equation in the system, the solution must lie on the graph of both equations. If the solution has integer values for both coordinates, a careful graph will yield the solution directly. Otherwise, the graphical solution will only give an approximate answer and indicate in which quadrant the solution lies. Any answer that you get using this method must be checked by substituting the values for $x$ and $y$ into both equations. If they are both satisfied then your answer is correct.

## Example 1

Solve the system of equations given by $\begin{cases} x+y=4 \\ 3x-y=0 \end{cases}$

**Solution:** Graph each line separately and look for the point of intersection. When graphing a line all you need is two points. I usually pick the intercepts whenever possible. If a line goes through the origin, then both the $x$- and $y$-intercepts are the same point. In that case I abandon my bias towards the intercepts and have to pick another point on the graph to get the second point. The $x$-intercept is the point (4, 0) and the $y$-intercept is the point (0, 4). Similarly, the $x$-intercept (and the $y$-intercept) of the line $3x - y = 0$ is the origin, or (0, 0). I will need to find a second point that lies on the line. If $x = 1$, then $y = 3$, so the point (1, 3) also lies on the line $3x - y = 0$. The graph of both of these lines is shown in Figure 8.1. Looking at the graph we can see that the intersection of these two points is the point (1, 3).

The last thing to do is verify that the point (1, 3) satisfies both equations. Check the first equation:

$x + y = 4$

$1 + 3 = 4$

Check the second equation:

$3x - y = 0$

$3 \cdot 1 - 3 = 0$

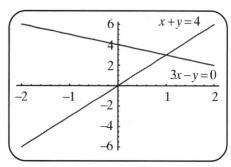

*Figure 8.1*

The point (1, 3) satisfies both equations. Therefore the point (1, 3) is a solution of the system of equations $\begin{cases} x + y = 4 \\ 3x - y = 0 \end{cases}$

## Lesson 8-1 Review

1. Solve the system of equations $\begin{cases} x - 2y = 0 \\ x + y = 10 \end{cases}$ by graphing each equation.

# Lesson 8-2: Solving Linear Systems by Substitution

The substitution method is an algebraic method for solving systems of equations. It involves solving for one of the variables in one of the equations in terms of the other one, and then substituting the expression for that variable into the other equation. In this process one of the variables is eliminated, making it possible to solve for the remaining one. I recommend using this method if at least one of the coefficients of one of the variables in one of the equations is either 1 or -1; in that case one variable is easy to isolate and you don't have to complicate the problem by having to work with fractions.

## Example 1

Solve the system of equations $\begin{cases} x + 2y = 5 \\ 2x - y = 5 \end{cases}$

**Solution:** Because the coefficient of $x$ in the top equation equals 1, the substitution method will work well. Use the first equation to solve for $x$:

$x + 2y = 5$

$x = -2y + 5$

I will refer to this equation as the **substitution equation**. Use this equation for $x$ to eliminate the appearance of $x$ in the second equation. In other words, replace every $x$ in the second equation with the expression $-2y + 5$. After that, solve for $y$:

$2x - y = 5$

$2(-2y + 5) - y = 5$

$-4y + 10 - y = 5$

$-5y + 10 = 5$

$-5y = -5$

$$\dfrac{\cancel{-5}y}{\cancel{-5}} = \dfrac{\cancel{-5}}{\cancel{-5}}$$

$y = 1$

Now that you know the value of $y$, use it in the substitution equation to solve for $x$:

$x = -2y + 5$

$x = -2(1) + 5$

$x = 3$

The last step is to check your work. Make sure that the point $(3, 1)$ satisfies both equations:

$$\begin{cases} x + 2y = 5 \\ 2x - y = 5 \end{cases}$$

$$\begin{cases} 3 + 2 \cdot 1 = 5 \\ 2 \cdot 3 - 1 = 5 \end{cases}$$

Both equations are satisfied, so the system of equations

$\begin{cases} x + 2y = 5 \\ 2x - y = 5 \end{cases}$ has solution $(3, 1)$.

## Example 2

Solve the system of equations: $\begin{cases} 3x - 2y = 1 \\ 4x - y = -2 \end{cases}$

**Solution:** Because the second equation has a variable with a coefficient of $-1$, we can use that equation to isolate $y$:

$4x - y = -2$

$-y = -4x - 2$

$(-1)(-y) = (-1)(-4x - 2)$

$y = 4x + 2$

Use this substitution equation to eliminate $y$ in the first equation and solve for $x$:

$3x - 2y = 1$

$3x - 2(4x + 2) = 1$

$3x - 8x - 4 = 1$

$-5x - 4 = 1$

$-5x = 5$

$$\frac{\cancel{-5}x}{\cancel{-5}} = \frac{\cancel{5}}{-\cancel{5}}$$

$x = -1$

Now that we know the value of $x$ we can use the substitution equation to solve for $y$:

$y = 4x + 2$

$y = 4(-1) + 2$

$y = -2$

So the solution consists of the point $(-1, -2)$. The last step is to check our solution to make sure that it satisfies both of the equations in our system:

$\begin{cases} 3x - 2y = 1 \\ 4x - y = -2 \end{cases}$

$\begin{cases} 3(-1) - 2(-2) = -3 + 4 = 1 \\ 4(-1) - (-2) = -4 + 2 = -2 \end{cases}$

Both of these equations are satisfied, so the system of equations

$\begin{cases} 3x - 2y = 1 \\ 4x - y = -2 \end{cases}$ has solution $(-1, -2)$.

The substitution method will work even if all of the coefficients of all of the variables in both of the equations are not 1 or $-1$. In that situation, I prefer to use the method that will be discussed in the next section.

## Lesson 8-2 Review

Use the substitution method to solve the following systems of equations:

1. $\begin{cases} 3x + y = 7 \\ x - 2y = 0 \end{cases}$    2. $\begin{cases} 2x + y = 5 \\ 2x - y = 0 \end{cases}$

# Lesson 8-3: Linear Systems and the Addition Method

This method is sometimes referred to as solving a linear system by using **linear combinations**. It should be used when all of the coefficients of the variables are real numbers other than 1 and $-1$. The procedure may be easier to understand with an example, so I will talk you through the logic and then walk you through a couple of examples so you can see how the technique is used. Then you will have a chance to apply what you have learned.

▶ To begin with, you should write each equation in *standard form* and write the variables in the same order. Line up your equations one on top of the other and decide which of the variables you want to eliminate.

▶ Look at the coefficient of that variable in each of the two equations. Your goal is to make those coefficients the same. The best way to do that is to find the least common multiple of the two coefficients. Figure out what you have to multiply by each equation in order to get those coefficients to be the same.

▶ Once you multiply each equation by its particular constant, you are ready to add or subtract the two equations, depending on the signs. If the two coefficients are the same sign, subtract one equation from the other; if the two coefficients are opposite in sign, add the sides of the two equations together. After you do that you will be left with only one variable.

▶ Solve for that variable and then use that value in either one of the original equations to solve for the value of the other variable. Let's see the method in action.

## Example 1

Solve the system of equations: $\begin{cases} 4x+3y=18 \\ 2x+5y=16 \end{cases}$

**Solution:** Each equation is already written in standard form. Notice that the coefficients of $x$ are 4 and 2; the coefficients of $y$ are 3 and 5. The least common multiple of 4 and 2 is 4; the least common multiple of 3 and 5 is 15. If I wanted to get rid of $x$ I would want the coefficient of $x$ to be 4 in both equations; if I wanted to eliminate $y$ I would want the coefficient of $y$ to be 15 in both equations. I will have to do less work if I eliminate $x$, so that is the variable I will go after. The way to have both equations have the same coefficient for $x$ is to multiply the second equation by 2 (and leave the first equation alone, because the coefficient of $x$ is already 4). When we multiply the second equation by 2, we will still have an equivalent equation, and hence an equivalent system of equations:

$\begin{cases} 4x+3y=18 \\ 4x+10y=32 \end{cases}$

Make sure that when you multiply an equation by a constant, you multiply each term in the equation by that constant. Now the system that we are solving is:

$\begin{cases} 4x+3y=18 \\ 4x+10y=32 \end{cases}$

Because the two coefficients in front of the variable $x$ are the same, we can subtract the bottom equation from the top equation:

$$\begin{array}{r} 4x+3y=18 \\ -\quad 4x+10y=32 \\ \hline 0x-7y=-14 \end{array}$$

Make sure that when you subtract the bottom equation, you subtract each and every term. You are using the distributive property: on the left-hand side of the equation you are evaluating $4x + 3y - (4x + 10y)$ and on the right-hand side of the equation you are evaluating $18 - 32$. Now we can solve directly for $y$ by dividing both sides of the equation by $-7$:

$$-7y = -14$$

$$\frac{\cancel{-7}y}{\cancel{-7}} = \frac{-14}{-7}$$

$$y = 2$$

Now that we know the value of $y$, we can use it to substitute into either of the original equations and solve for $x$. I will use the first equation to find $x$:

$$4x + 3y = 18$$

$$4x + 3 \cdot 2 = 18$$

$$4x + 6 = 18$$

$$4x = 12$$

$$\frac{\cancel{4}x}{\cancel{4}} = \frac{12}{4}$$

$$x = 3$$

So the solution to the system of equations is the point $(3, 2)$. The last thing to do is to check the solution in the second equation. Plug in the value of $y$ into the second equation:

$$4x + 10y = 32$$

$$4 \cdot 3 + 10 \cdot 2 = 32$$

So the point $(3, 2)$ satisfies both equations and is therefore the solution.

## Example 2

Solve the system of equations $\begin{cases} 2x - 3y = 0 \\ 3x - 2y = 5 \end{cases}$

**Solution:** Because none of the coefficients are equal to 1 or −1, the method to use would be the addition method. The numbers you pick to multiply the first and second equation by depend on which variable you want to eliminate. In order to eliminate $x$ from the system, you would need to multiply the first equation by 3 and the second equation by 2. If you wanted to eliminate $y$ from the system you would need to multiply the first equation by 2 and the second equation by 3. In the last example I eliminated $x$, so in this example I will eliminate $y$. First, I will transform both equations:

$$\begin{cases} 2x - 3y = 0 \\ 3x - 2y = 5 \end{cases}$$

Multiply the top equation by 2
Multiply the bottom equation by 3

$$\begin{cases} 4x - 6y = 0 \\ 9x - 6y = 15 \end{cases}$$

Now that the coefficients of $y$ are the same, I will carefully subtract the bottom equation from the top of the equation:

$$\begin{array}{r} 4x - 6y = 0 \\ -\ \ 9x - 6y = 15 \\ \hline -5x + 0y = -15 \end{array}$$

Finally, use the resulting equation to solve for $x$:

$$-5x = -15$$

$$\frac{\cancel{5}x}{\cancel{5}} = \frac{-15}{-5}$$

$$x = 3$$

Now that $x$ has been determined, use either of the original equations to solve for $y$. I will substitute the value of $x$ into the original second equation and solve for $y$:

$$3x - 2y = 5$$

$$3 \cdot 3 - 2y = 5$$

$$9 - 2y = 5$$

$$-2y = -4$$

$$y = 2$$

The solution to the system of equations is the point (3, 2). The last step is to check our answer by substituting this point into the first equation:

$$2x - 3y = 0$$

$$2 \cdot 3 - 3 \cdot 2 = 0$$

Because the point (3, 2) satisfies both equations, it is the solution to the system of equations.

Work out the following problems to be sure that you understand the technique used and can successfully apply it. Be sure to check your answers before moving on to the next lesson.

## Lesson 8-3 Review

Use the addition method to solve the following systems of equations:

1. $\begin{cases} 3x + 4y = 1 \\ 2x - 3y = -5 \end{cases}$
2. $\begin{cases} 2x + 3y = 3 \\ 3x - 2y = 11 \end{cases}$
3. $\begin{cases} 3x + 5y = -1 \\ 4x + 3y = -5 \end{cases}$

# Lesson 8-4: Linear Systems With Three Variables

The techniques used to solve systems of equations with two variables can be adapted to solve systems of equations with three or more variables. A linear equation with three variables, $x$, $y$, and $z$, is an equation of the form $Ax + By + Cz = D$, where $A$, $B$, $C$, and $D$ are real numbers. The graph of a linear equation with three variables is actually a plane in three-dimensional space, and graphing it on a two-dimensional sheet of paper requires some imagination: imagine a sheet of paper (as a model of a plane) floating in space, and extending in all directions. It is important to realize that even though the equation $Ax + By + Cz = D$ is called a linear equation, its graph is a plane in space. Because the graph of $Ax + By + Cz = D$ is difficult to draw on a sheet of paper, using a graphical approach to solve systems of equations with three variables can be awkward. Fortunately, the algebraic methods for solving systems of equations with two variables will carry over to the situation in which there are three (or more) variables involved.

As we saw with linear systems with two variables, a linear system of equations with three or more variables can have one, none, or infinitely many solutions. The concepts of consistent, inconsistent, dependent, and independent systems applies to all systems of linear equations, regardless of the number of variables.

You can use either the substitution method or the addition method to solve systems of linear equations regardless of the number of variables. Use the substitution method if one of the coefficients of one of the variables in one of the equations equals 1.

## Example 1

Solve the system of equations $\begin{cases} x - 3y - 8z = -12 \\ 2x + 3y + 4z = 5 \\ 3x - y + 4z = 0 \end{cases}$

**Solution:** The variable $x$ has a coefficient of 1 in the first equation, so we can use that equation to write $x$ in terms of $y$ and $z$, and knock out $x$ in the second and third equations. We will then end up with a system of two equations and two unknowns, which we can solve to determine $y$ and $z$. We can use the values for $y$ and $z$ to find $x$. First, use the first equation to write $x$ in terms of $y$ and $z$:

$x = 3y + 8z - 12$.

Next, substitute in for $x$ in the other two equations and simplify:

Substitute in for $x$: $\begin{cases} 2(3y + 8z - 12) + 3y + 4z = 5 \\ 3(3y + 8z - 12) - y + 4z = 0 \end{cases}$

Distribute and collect terms: $\begin{cases} 6y + 16z - 24 + 3y + 4z = 5 \\ 9y + 24z - 36 - y + 4z = 0 \end{cases}$

$\begin{cases} 9y + 20z = 29 \\ 8y + 28z = 36 \end{cases}$

Divide each term in the second equation by 2: $\begin{cases} 9y + 20z = 29 \\ 4y + 14z = 18 \end{cases}$

At this point, we can use the addition method to solve this new (more manageable) system of equations that only involves two variables.

Multiply the top equation by 4 and the bottom equation by 9:

$\begin{cases} 36y + 80z = 116 \\ 36y + 126z = 162 \end{cases}$

Subtracting one equation from the other, we see that $z = 1$. From this, we can find $y$: $y = 1$.

Substituting in to find $x$, we have $x = 3 + 8 - 12 = -1$. The solution to this system of equations is the point $(-1, 1, 1)$.

## Lesson 8-4 Review

Solve the following systems of equations:

1. $\begin{cases} x + 2y - 3z = 0 \\ 2x + 3y + z = -1 \\ x + 4y + 2z = 2 \end{cases}$
  
2. $\begin{cases} x + y + z = 3 \\ 2x + y + 3z = 2 \\ x + y - z = 5 \end{cases}$

# Lesson 8-5: Applications

As we will discuss in this lesson, systems of linear equations can be used to solve a variety of applications. For example, finding the equation of the line passing through the points $(1, 3)$ and $(-1, 4)$ is equivalent to

solving the system of equations $\begin{cases} m + b = 3 \\ -m + b = 4 \end{cases}$. To understand this, remember

that the slope-intercept equation of a line is $y = mx + b$. If we substitute $x = 1$ and $y = 3$ (which are the coordinates of the first point) into this equation, the result is $m + b = 3$. If we substitute $x = -1$ and $y = 4$ (which are the coordinates of the second point) into this equation, we have $-m + b = 4$.

These are the two equations in the system $\begin{cases} m + b = 3 \\ -m + b = 4 \end{cases}$.

We can solve this system by either the substitution method, or the addition method. Because the signs of the coefficients of $m$ are opposites of each other, adding these two equations and dividing by 2 will immediately reveal the value of $b$, which is the $y$-intercept of the line: $b = \frac{7}{2}$. Subtracting one equation from the other and dividing by 2 will reveal the value of $m$, which is the slope of the line: $m = -\frac{1}{2}$. The equation of the line passing through the points $(1, 3)$ and $(-1, 4)$ is $y = -\frac{1}{2}x + \frac{7}{2}$.

A linear function is a function of the form $f(x) = mx + b$. There are two constants that need to be determined: $m$ and $b$. Given two points that satisfy the linear function, we can generate a system of two equations and two unknowns (each point generates one of the two equations). A parabola is a function of the form $f(x) = ax^2 + bx + c$. In order to define a parabola, we need to know the values of three constants: $a$, $b$, and $c$. We will need three points to generate a system of three equations and three unknowns.

## Example 1

Find the equation of the parabola that passes through the points $(0, -4)$, $(1, 1)$, and $(-1, -5)$.

**Solution:** The general formula for a parabola is $f(x) = ax^2 + bx + c$. This parabola must pass through the points $(0, -4)$, $(1, 1)$, and $(-1, -5)$. Evaluate the quadratic function at each of the three points to generate a system of equations:

$$\begin{cases} 0 \cdot a + 0 \cdot b + c = -4 \\ 1^2 \cdot a + 1 \cdot b + c = 1 \\ (-1)^2 \cdot a + (-1) \cdot b + c = -5 \end{cases}$$

The first equation is the result when substituting $f(0) = -4$ ($x = 0$ and $f(0) = -4$), the second equation is the result when using the second point $f(1) = 1$, and the third equation comes from using the point $f(-1) = -5$. To solve this system of equations, notice that the first equation yields the value of $c$ immediately: $c = -4$. We can substitute this value of $c$ into the other two equations:

$$\begin{cases} a + b + -4 = 1 \\ a - b - 4 = -5 \end{cases}$$

$$\begin{cases} a + b = 5 \\ a - b = -1 \end{cases}$$

This system of equations can be solved by the addition method (or the substitution method): $a = 2$ and $b = 3$. The equation of the parabola is $f(x) = 2x^2 + 3x - 4$.

We can also use systems of equations to find the equation of a plane. If the equation of a plane is written $z = Ax + By + C$, there are three constants that need to be determined: $A$, $B$, and $C$. You may recall from geometry that there is only one plane that contains a set of three non-collinear points. Given the $x$, $y$, and $z$ coordinates of three points, we can substitute that information into the formula $z = Ax + By + C$ to generate a system of three equations and three unknowns. The solution to that system of equations are the coefficients $A$, $B$, and $C$ that appear in the formula.

## Example 2

Find the equation of the plane that passes through the points $(1, 1, 2)$, $(-1, 2, -3)$, and $(0, 2, 0)$.

**Solution:** The equation of a plane can be written $z = Ax + By + C$. Substitute in for $x, y$, and $z$ for each of the three points to generate a system of equations:

$$\begin{cases} A \cdot 1 + B \cdot 1 + C = 2 \\ A \cdot (-1) + B \cdot 2 + C = -3 \\ A \cdot 0 + B \cdot 2 + C = 0 \end{cases}$$

$$\begin{cases} A + B + C = 2 \\ -A + 2B + C = -3 \\ 2B + C = 0 \end{cases}$$

Use the third equation to solve for $C$ in terms of $B$, and reduce the system to two unknowns:

$$C = -2B$$

$$\begin{cases} A + B - 2B = 2 \\ -A + 2B - 2B = -3 \end{cases}$$

$$\begin{cases} A - B = 2 \\ -A = -3 \end{cases}$$

From this, we see that $A = 3$, $B = 1$, and $C = -2$. The equation of the plane is $z = 3x + y - 2$.

In addition to being used to find equations of basic functions, systems of equations can be used to model various situations. Systems of equations can be used to solve mixture problems, rate problems, and financial problems. These problems are usually presented in the form of a word problem. There are two aspects of solving a word problem:

▶ Translating the given information into a system of equations.

▶ Solving the system of equations.

A mistake in either of the two steps results in an incorrect solution, so it is important to read a word problem carefully to create the appropriate system of equations.

## Example 3

A smooth breakfast blend of coffee is made by mixing Columbian coffee, costing $8 per pound, with Sumatran coffee that costs $15 per pound. How many pounds of each type of coffee are required to make 70 pounds of the breakfast blend if the cost of the breakfast blend is $10.80 per pound?

**Solution:** With these types of problems, we will generate two equations by conserving coffee and money. Let $c$ represent the number of pounds of Columbian coffee, and $s$ denote the number of pounds of Sumatran coffee. Conservation of coffee gives the equation $c + s = 70$. We also need to conserve money. Every pound of Columbian coffee is equivalent to $8, so the total cost of the Columbian coffee is $8c$. Every pound of Sumatran coffee is equivalent to $15, so the total cost of the Sumatran coffee is $15s$. Every pound of the mixture costs $10.80, and we have 70 pounds of the mixture, so the total cost of the mixture is $(10.8)(70)$, or $756. Also, the total cost of the mixture is the total cost of the Columbian coffee plus the total cost of the Sumatran coffee. Conservation of money gives the equation $8c + 15s = 756$. Our system of equations is:

$$\begin{cases} c + s = 70 \\ 8c + 15s = 756 \end{cases}$$

and we can solve this system using the substitution method:

$c = 70 - s$

$8(70 - s) + 15s = 756$

$560 - 8s + 15s = 756$

$7s = 196$

$s = 28$

We can use the substitution equation to solve for $c$: $c = 70 - 28 = 42$. The mixture has 42 pounds of Columbian coffee and 28 pounds of Sumatran coffee.

## Example 4

A plane flies 560 miles in 1.75 hours traveling with the wind. The return trip against the wind takes the plane 2 hours. Find the speed of the plane and the speed of the wind.

**Solution:** When solving rate problems, I recommend creating a chart for the rate, time, and distance traveled. Let $p$ represent the speed of the plane, and $w$ represent the speed of the wind. When traveling with the wind, the speed of the plane and the speed of the wind are added together to give the overall speed for the trip. When traveling against the wind, the speed of the wind is subtracted from the speed of the plane to give the overall speed for the trip.

| Trip | Rate | Time | Distance |
|------|------|------|----------|
| With the wind | $p + w$ | 1.75 | 560 |
| Against the wind | $p - w$ | 2.00 | 560 |

We can use the equation rate × time = distance to generate the system of equations:

$$\begin{cases} 1.75(p+w) = 560 \\ 2(p-w) = 560 \end{cases}$$

If we divide both sides of the first equation by 1.75, and both sides of the second equation by 2, our system of equations will be easy to solve:

$$\begin{cases} p+w = 320 \\ p-w = 280 \end{cases}$$

Using the addition method, adding the equations gives $p$: $p = 300$. Subtracting one equation from the other gives $w$: $w = 20$. The speed of the plane is 300 mph, and the speed of the wind is 20 mph.

## Example 5

A value meal consists of a cheeseburger, fries, and a soda. Last year, a value meal cost $4.35 Over the past year, the price of a cheeseburger increased by 20%, the price of fries increased by 10%, and the price of a soda increased by 50%. If the value meal now costs $5.40, and the fries cost as much as a soda, how much does a cheeseburger cost now?

**Solution:** Let $b$ denote the price of the burger, $f$ denote the price of the fries, and $c$ represent the price of a soda *last year*. Then our first equation is $b + f + c = 4.35$. Taking into consideration the increase in prices, we can create the second equation: $1.20b + 1.10f + 1.5c = 5.40$.

We also know that the current price of the fries equals the current price of the soda: $1.10f = 1.5c$. Our system of equations is:

$$\begin{cases} b + f + c = 4.35 \\ 1.20b + 1.1f + 1.5c = 5.40 \\ 1.1f = 1.5c \end{cases}$$

If we multiply the first equation by 1.1, we will be able to replace $1.1f$ with $1.5c$ in the first two equations:

$$\begin{cases} 1.1b + 1.1f + 1.1c = 4.785 \\ 1.20b + 1.1f + 1.5c = 5.40 \end{cases}$$

$$\begin{cases} 1.1b + 1.5c + 1.1c = 4.785 \\ 1.20b + 1.5c + 1.5c = 5.40 \end{cases}$$

$$\begin{cases} 1.1b + 2.6c = 4.785 \\ 1.20b + 3c = 5.40 \end{cases}$$

Now use the addition method: multiply the first equation by 3 and the second equation by 2.6:

$$\begin{cases} 3.3b + 7.8c = 14.355 \\ 3.12b + 7.8c = 14.04 \end{cases}$$

Subtract one equation from the other:

$$0.18b = 0.315$$

Now divide both sides of the equation by 0.18 to solve for $b$:

$$b = 1.75$$

The price of the cheeseburger last year was \$1.75. The price went up 20%, so the current price is:

$$1.75 \cdot 1.20 = 2.10$$

The current price of the cheeseburger is \$2.10.

---

## Lesson 8-5 Review

1. Find the equation of the parabola that passes through the points $(0, 4)$, $(1, 5)$, and $(-1, 9)$.

2. Find the equation of the plane that passes through the points $(1, 1, 5)$, $(0, 1, 1)$, and $(-1, 0, -1)$.

# Lesson 8-6: Solving Non-Linear Systems

We began our study of functions by studying the properties of linear functions. Linear functions are extremely nice functions because they are so easy to understand. Finding the equation of a linear function only requires two points. Once you know two points that satisfy a linear function, you know everything there is to know about that function. If linear functions were the only functions in town, our work would be done. Fortunately, there are polynomial, exponential, and logarithmic functions, just to name a few.

When we solve systems of *linear* equations, there are only three possible outcomes: the system of equations has no solution, one solution, or infinitely many solutions. Non-linear systems of equations do *not* have this property. Non-linear systems of equations can have *any* number of solutions.

Non-linear systems of equations can be very complicated. It is for this reason that mathematicians spend a significant amount of time finding linear approximations to non-linear functions. In this lesson, we will work with some familiar non-linear functions, such as parabolas and the absolute value function. We will also work with circles, which are not even functions!

Non-linear systems of equations can be solved using the substitution method, the graphical method, and the elimination method. Some systems of equations require you to be imaginative, and to use every tool at your disposal. You may need to multiply both sides of an equation by a variable, square both sides of an equation, or carefully remove absolute value symbols.

## Example 1

Solve the system of equations: $\begin{cases} y = x^2 - 3x \\ y = x - 3 \end{cases}$

**Solution:** This system of equations involves a parabola and a line. We can graph these two equations and find their intersection. The graphs of these two functions are shown in Figure 8.2.

The graphical method should only be used to determine the number of solutions and estimate the points of intersection. From the graphs of these two functions, we expect to find two solutions to this system of equations. To find the exact values

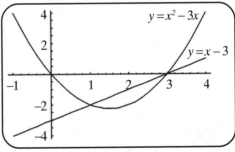

*Figure 8.2*

of the solutions we will need to solve this system algebraically. The first step in solving non-linear systems of equations is to look at the equations involved. Notice that $y$ is isolated in each of the equations in this system. The easiest way to solve this system of equations is to use the substitution method and solve for $x$. When we do this, we will end up with a quadratic equation in $x$, which we can solve by either factoring or using the quadratic formula:

Set the two equations equal to each other.    $x^2 - 3x = x - 3$

Collect like terms.    $x^2 - 4x + 3 = 0$

Factor the quadratic expression.    $(x - 3)(x - 1) = 0$

Set each factor equal to 0 and solve.    $x = 3 \text{ or } x = 1$

Each of these two values of $x$ will result in a value for $y$. We can substitute the value of $x$ into either (or both) equations.
When $x = 3$, $y = 0$, and when $x = 1$, $y = -2$. The solutions to this system of equations are $(3, 0)$ and $(1, -2)$.

## Example 2

Solve the system of equations $\begin{cases} y = |3x| \\ y = x^2 + 2 \end{cases}$

**Solution:** The first function involves an absolute value, and the second function is a parabola that opens up and has a vertex at $(0, 2)$. Both of these functions are even functions, and their graphs are symmetric with respect to the y-axis. The graphs of these two functions are shown in Figure 8.3.

From the graphs of these two functions, we see that there are several points of intersection. To find the solutions to this system of equations, we will need to set the two functions of $x$ equal to each other and then carefully remove the absolute value symbols to create two systems of equa-tions. We can use the sym-

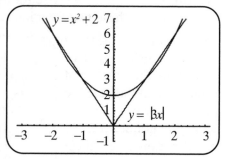

*Figure 8.3*

metry of the two equations to our advantage: if $(a, b)$ is a solution to this system of equations, so will $(-a, b)$ be. We can use the substitution method to solve this system of equations. We will then remove the absolute value symbols and create two new equations:

$$x^2 + 2 = |3x|$$

| | |
|---|---|
| $x^2 + 2 = 3x$ | $x^2 + 2 = -3x$ |
| $x^2 - 3x + 2 = 0$ | $x^2 + 3x + 2 = 0$ |
| $(x - 2)(x - 1) = 0$ | $(x + 2)(x + 1) = 0$ |
| $x = 2$ or $x = 1$ | $x = -2$ or $x = -1$ |

Notice the symmetry of the two sets of values of $x$. This is what we expected when we first looked at the system of equations. To find the solutions to the system of equations, we need to substitute our values of $x$ into the original system of equations and find $y$. When $x = 2$, $y = 6$, and when $x = -2$, $y = 6$. When $x = 1$, $y = 3$, and when $x = -1$, $y = 3$. There are four solutions to this system of equations: $\{(2, 6), (-2, 6), (1, 3), (-1, 3)\}$.

## Example 3

Solve the system of equations:
$$\begin{cases} \dfrac{2}{x}+\dfrac{1}{y}=2 \\ \dfrac{4}{x}+\dfrac{1}{y}=3 \end{cases}$$

**Solution:** We can subtract one equation from the other and cancel out the terms that involve $y$:

$$\frac{2}{x}+\frac{1}{y}=2$$

$$-\ \ \frac{4}{x}+\frac{1}{y}=3$$

$$\overline{\quad -\frac{2}{x}=-1\quad}$$

$$x=2$$

Substitute this value into either of the equations in the system:

$$\frac{2}{2}+\frac{1}{y}=2$$

$$y=1$$

The solution to this system of equations is (2, 1).

## Lesson 8-6 Review

Solve the following systems of equations:

1. $\begin{cases} 5x-y=6 \\ y=x^2 \end{cases}$
2. $\begin{cases} y=\sqrt{x+3} \\ x+y=3 \end{cases}$
3. $\begin{cases} x^2+y^2=5 \\ x-y=1 \end{cases}$

# Lesson 8-7: Linear Inequalities

Linear inequalities can be used in business to help with the problem of allocating resources. For example, a company may have an advertising budget that can be spent on television ads, magazine ads, and internet ads. Systems of linear inequalities can help a company decide how to spend its advertising dollars to efficiently reach its target audience. Similarly, a company that builds microprocessors has to decide how many of each type of computer chip to make so that it maintains its market share and stays competitive. Systems of linear inequalities can help a company decide how to allocate their resources and produce an optimal amount of each type of computer chip. The airline industry can use systems of linear equations to help determine how to schedule its planes so that they serve the greatest number of people. The process of using linear inequalities in the decision-making process is called **linear programming**. In this lesson, we will learn how to solve linear inequalities and systems of linear inequalities.

The graph of the equation $x + 2y = 4$ is a line with slope $-\frac{1}{2}$ and y-intercept $(0, 2)$. We say that the solution to this equality consists of the set of all points that lie on this line. The equation $x + 2y > 4$ is a region in the coordinate plane that is made of all points that satisfy the inequality. The graph of this region is shown in Figure 8.4. I will walk you through the process of graphing this linear inequality.

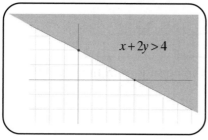

*Figure 8.4*

To graph a linear inequality, first pretend that you have a linear equality. Rewrite the problem and replace the inequality with an equality: $x + 2y = 4$. This is the equation of a line, and all we need to graph the equation of a line are two points. The two intercepts of a line usually work fine. The intercepts of the line $x + 2y = 4$ are $(0, 2)$ and $(4, 0)$. Plot these two points and get ready to draw the line connecting them. Before you connect the two points, however, we need to determine whether to draw a solid line or a dashed line, and the original inequality is what dictates that. A solid line is drawn when the original inequality is $\geq$ or $\leq$, and a dashed line is drawn when the original inequality is $>$ or $<$, or is a strict inequality. In our case, the original inequality was a strict inequality, so we will need to connect our two points using a dashed line. This line divides the plane into two regions. One region will satisfy the inequality, the other will not. The easiest way to determine which region works is to choose a test point in one of the regions. Evaluate the inequality at the test value and check whether or not the inequality is satisfied. If it is, then you can conclude that every point in that region will satisfy the inequality. If the inequality is not satisfied, then you can conclude that every point in the *other* region will satisfy the inequality. For the inequality $x + 2y > 4$, the test point that I will use is the origin: If I substitute $x = 0$ and $y = 0$ into the inequality, the result is the statement $0 > 4$. This is clearly ludicrous, so the side of the line that contains the origin does not satisfy the inequality; the opposite side does. That's the general process for solving a linear inequality.

## Example 1

Graph the solutions to the inequality $2x - 3y \leq 6$.

**Solution:** The first step is to graph the line $2x - 3y = 6$. The intercepts of this line are the points $(3, 0)$ and $(0, -2)$. Before we connect these two points, we need to decide whether to use a solid or a dashed line. Because the inequality is $\leq$, we will use a solid line. Finally, determine which side of the line satisfies the inequality. The test value that I will use is the origin: substituting $x = 0$ and $y = 0$ into the inequality yields the statement $0 \leq 6$. This statement is true, so the region containing the origin satisfies the inequality. The solution to this inequality is shown in Figure 8.5.

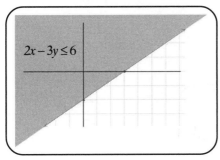

*Figure 8.5*

A **system of inequalities** is a collection of inequalities. Solving a system of inequalities involves graphing the solution to each inequality and looking for the region where they overlap.

## Example 2

Solve the following system of inequalities: $\begin{cases} x - 2y \leq 2 \\ x + y > 3 \end{cases}$

**Solution:** Graph each inequality and look for the region where they overlap. The solution to the inequality is shown in Figure 8.6

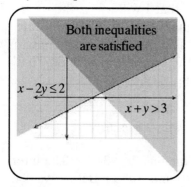

*Figure 8.6*

## Lesson 8-7 Review

1.  Solve $\begin{cases} x - 3y \le 6 \\ 2x + y > 2 \end{cases}$

## Answer Key

### Lesson 8-1 Review

1.  The solution is $(4, 2)$.

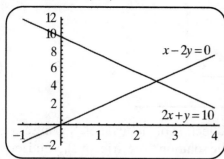

*Figure 8.7*

### Lesson 8-2 Review

1.  $(2, 1)$

2.  $\left( \frac{5}{4}, \frac{5}{2} \right)$

## Lesson 8-3 Review

1. $(-1, 1)$        2. $(3, -1)$        3. $(-2, 1)$

## Lesson 8-4 Review

1. $(-2, 1, 0)$        2. $(1, 3, -1)$

## Lesson 8-5 Review

1. $f(x) = 3x^2 - 2x + 4$      2. $z = 4x - 2y + 3$

## Lesson 8-6 Review

1. $(3, 9)$ and $(2, 4)$

2. $(1, 2)$ ($y = -3$ leads to an extraneous solution)

3. $(-1, -2)$ and $(2, 1)$

## Lesson 8-7 Review

1.

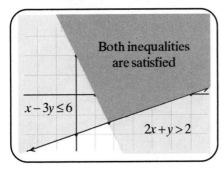

*Figure 8.8*

# Matrices

The techniques that we used to solve systems of equations (substitution and addition) work well for systems of two equations and two unknowns. Applying those same techniques to solve systems that involve three equations and three unknowns is more tedious. Imagine using those techniques to solve systems of equations that involve hundreds, or even thousands, of equations and unknowns! Solving large systems of equations requires a significant amount of computational power. The substitution and elimination methods are easy to understand, but are problematic when trying to solve them using a computer. In this chapter we will learn about matrices, and discover how to use matrices to solve systems of linear equations.

## Lesson 9-1: The Algebra of Matrices

A **matrix** is a rectangular array of real numbers. The rows of a matrix run horizontally and the columns run vertically. A matrix with only one row is called a **row matrix**, and a matrix with only one column is called a **column matrix**. The size, or dimension, of a matrix is determined by the number of rows and the number of columns. A matrix with $m$ rows and $n$ columns is an $m \times n$ (pronounced "m by n") matrix. The size is always given by the number of rows by the number of columns. When I first learned about matrices, I also drank RC cola, which helped me remember that the rows come first when giving the dimensions of a matrix. For example, the matrix $\begin{bmatrix} 3 & -1 & 2 \\ 4 & 0 & 8 \end{bmatrix}$ has two rows and three columns, so it is a $2 \times 3$ matrix.

The matrix $\begin{bmatrix} 1 & 2 \\ 3 & 4 \end{bmatrix}$ is an example of a **square** matrix, or a matrix with the same number of rows and columns. Each number in a matrix is called an **element** of the matrix. An element of a matrix is denoted by $a_{ij}$, where $i$ denotes the row of the element and $j$ represents the column of the element. In the matrix $\begin{bmatrix} 3 & -1 & 2 \\ 4 & 0 & 8 \end{bmatrix}$, $a_{23} = 8$ and $a_{12} = -1$. Again, the row of the element is written first, followed by the column of the element.

The algebra of matrices refers to the rules established for combining matrices. Matrices can be added, subtracted, and multiplied. Actually, there are two types of matrix multiplication.

Two matrices can be added or subtracted only if they have the same dimension. The process by which matrices are added or subtracted is the most natural one that comes to mind: addition and subtraction are performed component-wise. In other words, when adding two matrices, just add the corresponding elements together. The reason that the dimensions have to be the same is that we don't want any of the elements to be left out. If two matrices are to be added, every element in one matrix has to have a corresponding element in the other matrix. Subtraction works the same way: subtraction is performed component-wise.

## Example 1

If $A = \begin{bmatrix} 1 & 2 \\ 3 & 4 \end{bmatrix}$, $B = \begin{bmatrix} 2 & 4 \\ 6 & 8 \end{bmatrix}$ and $C = \begin{bmatrix} 4 & 1 & 0 \\ 3 & 2 & -1 \end{bmatrix}$, find the following:

a. $A + B$        b. $A + C$        c. $A - B$

**Solution:**

a. $A + B = \begin{bmatrix} 1 & 2 \\ 3 & 4 \end{bmatrix} + \begin{bmatrix} 2 & 4 \\ 6 & 8 \end{bmatrix} = \begin{bmatrix} 3 & 6 \\ 9 & 12 \end{bmatrix}$

b. $A + C$ is not possible, because they do not have the same dimensions.

c. $A - B = \begin{bmatrix} 1 & 2 \\ 3 & 4 \end{bmatrix} - \begin{bmatrix} 2 & 4 \\ 6 & 8 \end{bmatrix} = \begin{bmatrix} -1 & -2 \\ -3 & -4 \end{bmatrix}$

Multiplication was originally used as a shorthand notation for repeated addition. For example, we can write $x + x + x + x$ more compactly as $4x$. The multiplication table involves memorizing this repeated addition for single digit numbers so that we perform this repetitive addition more quickly. Rather than add 8 to itself 10 times, we write $8 \cdot 10$ and are familiar with multiplication enough to know that the result is 80. Repeated addition of matrices will be analogous. Instead of writing

$\begin{bmatrix} 1 & 2 \\ 3 & 4 \end{bmatrix} + \begin{bmatrix} 1 & 2 \\ 3 & 4 \end{bmatrix} + \begin{bmatrix} 1 & 2 \\ 3 & 4 \end{bmatrix} + \begin{bmatrix} 1 & 2 \\ 3 & 4 \end{bmatrix} + \begin{bmatrix} 1 & 2 \\ 3 & 4 \end{bmatrix}$, we will write $5 \cdot \begin{bmatrix} 1 & 2 \\ 3 & 4 \end{bmatrix}$. This is

called multiplying a matrix by a **scalar**, which is just another word for a *number*. Scalar multiplication is done component-wise, in the sense that every element in the matrix will be multiplied by 5:

$$5 \cdot \begin{bmatrix} 1 & 2 \\ 3 & 4 \end{bmatrix} = \begin{bmatrix} 5 \cdot 1 & 5 \cdot 2 \\ 5 \cdot 3 & 5 \cdot 4 \end{bmatrix} = \begin{bmatrix} 5 & 10 \\ 15 & 20 \end{bmatrix}.$$

Most of the operations that we do with matrices are analogous to the operations with real numbers. In the real number system, there is an additive identity, called 0, that has the property that adding it to any real number doesn't change the number. The *zero matrix* is a matrix with that same property. Whenever you add the zero matrix to any matrix (with the proper dimension, of course), you don't change the matrix. It should not surprise you to learn that all of the elements in the zero matrix are 0.

Every real number has an additive inverse. The additive inverse of 3 is $-3$, and the additive inverse of $-6$ is 6. The sum of a number and its additive inverse is 0. Matrices also have this property. The sum of a matrix and its additive inverse is the *zero matrix*. For example, the additive inverse

of $\begin{bmatrix} 1 & 2 \\ 3 & 4 \end{bmatrix}$ is $\begin{bmatrix} -1 & -2 \\ -3 & -4 \end{bmatrix}$.

## Lesson 9-1 Review

1. If $A = \begin{bmatrix} 2 & -3 \\ 1 & 4 \end{bmatrix}$, $B = \begin{bmatrix} 8 & 2 \\ -1 & 8 \end{bmatrix}$ and $C = \begin{bmatrix} 2 & 5 & 0 \\ -9 & 3 & -1 \end{bmatrix}$, determine:

    a. $A + B$                    c. $5C$

    b. $3A - B$               d. The additive inverse of $C$

# Lesson 9-2: Matrix Multiplication

There are times when two matrices can be multiplied together. The definition of matrix multiplication is a little strange. Your idea of matrix multiplication may involve multiplication component-wise, but that would be too easy. It turns out that, in the grand scheme of matrix algebra, component-wise multiplication does not have the properties that we want matrix multiplication to have. If you want to learn more about why matrix multiplication is defined the way it is, you will need to take a class in linear algebra. For now, you will have to trust me that this strange method of matrix multiplication really is the best option available.

When you multiply two real numbers together, the order in which you multiply them doesn't matter: $2 \cdot 6$ and $6 \cdot 2$ both are equal to 12. Multiplication of real numbers commutes. Unfortunately, matrix multiplication does not have this property. Matrix multiplication is *not* commutative: if $A$ and $B$ are matrices, then $A \cdot B$ and $B \cdot A$ are usually different. The order in which you multiply two matrices matters. Matrix multiplication is also complicated by the fact that not all matrices can be multiplied. So, before you try to multiply two matrices, the first thing to do is check to see if the two matrices can, in fact, be multiplied together. If $A$ and $B$ are matrices, then $A \cdot B$ can be found only if each row of $A$ has the same number of elements as each column of $B$. If A is an $m \times n$ matrix, and if $B$ is an $n \times p$ matrix, then it will be possible to find $A \cdot B$; $A \cdot B$ makes sense if the number of columns of $A$ equal the number of rows of $B$. If you write the dimensions of each matrix next to each other, $(m \times n)(n \times p)$, notice that the middle letters match up. That is a quick way to see if two matrices can be multiplied. For example, if $A$ is a $3 \times 4$ matrix, and if $B$ is a $4 \times 10$ matrix, then we can find the product $A \cdot B$ because, if we line up the dimensions of the two matrices in the order in which we will multiply them, we have $(3 \times 4)(4 \times 10)$. Notice that the product $B \cdot A$ is not possible. Write the dimensions of $B$ and $A$ in the order of the product $B \cdot A$: $(4 \times 10)(3 \times 4)$. The inner numbers are not the same, so the product $B \cdot A$ cannot be done.

When you line up the dimensions of the two matrices, checking the inner two values determines whether the multiplication is possible. The outer values give the dimensions of the resulting product. For example, if $A$ is a $3 \times 4$ matrix, and if B is a $4 \times 10$ matrix, then the dimension of the product $A \cdot B$ will be $3 \times 10$. Line up the dimensions of $A$ and $B$ and read off the outer values. The outer values of $(3 \times 4)(4 \times 10)$ give $3 \times 10$.

If two matrices, $A$ and $B$, can be multiplied together, then the element $c_{ij}$ is found by combining the $i$th row of $A$ with the $j$th column of $B$ component-wise, using the formula

$$c_{ij} = a_{i1}b_{1j} + a_{i2}b_{2j} + ... + a_{in}b_{nj}$$

It's probably easier to see this matrix multiplication in action.

## Example 1

Find the product $A \cdot B$ if $A = \begin{bmatrix} 1 & 3 \\ 7 & 9 \end{bmatrix}$ and $B = \begin{bmatrix} -2 \\ 4 \end{bmatrix}$.

**Solution:** First, check to see if we can find this product. $A$ is a $2 \times 2$ matrix, and $B$ is a $2 \times 1$. If we line up our dimensions, we have $(2 \times 2)(2 \times 1)$. The inner numbers are the same, so we can find this product. The dimensions of the resulting matrix will be $2 \times 1$. Now let's do the multiplication. First, we will take the first row of $A$ with the first column of $B$:

$$\begin{bmatrix} 1 & 3 \\ 7 & 9 \end{bmatrix} \cdot \begin{bmatrix} -2 \\ 4 \end{bmatrix} = \begin{bmatrix} (1)(-2)+(3)(4) \end{bmatrix} = \begin{bmatrix} 10 \end{bmatrix}.$$

Next, take the second row of $A$ with the first column of $B$:

$$\begin{bmatrix} 1 & 3 \\ 7 & 9 \end{bmatrix} \cdot \begin{bmatrix} -2 \\ 4 \end{bmatrix} = \begin{bmatrix} 10 \\ (7)(-2)+(9)(4) \end{bmatrix} = \begin{bmatrix} 10 \\ 22 \end{bmatrix}.$$

The product is $\begin{bmatrix} 1 & 3 \\ 7 & 9 \end{bmatrix} \cdot \begin{bmatrix} -2 \\ 4 \end{bmatrix} = \begin{bmatrix} 10 \\ 22 \end{bmatrix}$.

## Example 2

Find the product $A \cdot B$ if $A = \begin{bmatrix} 1 & 3 \\ 2 & -1 \end{bmatrix}$ and $B = \begin{bmatrix} -1 & 2 \\ 4 & 0 \end{bmatrix}$.

**Solution:** First, verify that the multiplication can be done. Because both A and B are square $2 \times 2$ matrices, we have the correct number of rows and columns to line everything up. The result will also be a $2 \times 2$ matrix. Work out each element in the product by taking rows of $A$ with columns of $B$:

$$\begin{bmatrix} 1 & 3 \\ 2 & -1 \end{bmatrix} \cdot \begin{bmatrix} -1 & 2 \\ 4 & 0 \end{bmatrix} = \begin{bmatrix} (1)(-1)+(3)(4) & \\ & \end{bmatrix} = \begin{bmatrix} 11 & \\ & \end{bmatrix}$$

$$\begin{bmatrix} 1 & 3 \\ 2 & -1 \end{bmatrix} \cdot \begin{bmatrix} -1 & 2 \\ 4 & 0 \end{bmatrix} = \begin{bmatrix} 11 & (1)(2)+(3)(0) \\ & \end{bmatrix} = \begin{bmatrix} 11 & 3 \\ & \end{bmatrix}$$

$$\begin{bmatrix} 1 & 3 \\ 2 & -1 \end{bmatrix} \cdot \begin{bmatrix} -1 & 2 \\ 4 & 0 \end{bmatrix} = \begin{bmatrix} 11 & 3 \\ (2)(-1)+(-1)(4) & \end{bmatrix} = \begin{bmatrix} 11 & 3 \\ -6 & \end{bmatrix}$$

$$\begin{bmatrix} 1 & 3 \\ 2 & -1 \end{bmatrix} \cdot \begin{bmatrix} -1 & 2 \\ 4 & 0 \end{bmatrix} = \begin{bmatrix} 11 & 3 \\ -6 & (2)(2)+(-1)(0) \end{bmatrix} = \begin{bmatrix} 11 & 3 \\ -6 & 4 \end{bmatrix}$$

The product is $\begin{bmatrix} 1 & 3 \\ 2 & -1 \end{bmatrix} \cdot \begin{bmatrix} -1 & 2 \\ 4 & 0 \end{bmatrix} = \begin{bmatrix} 11 & 3 \\ -6 & 4 \end{bmatrix}.$

---

Two matrices are equal if they have the same dimension and all pairs of corresponding elements are equal. For example, if the two matrices $\begin{bmatrix} 2 & -1 \\ 3 & x \end{bmatrix}$ and $\begin{bmatrix} 2 & y \\ z & 4 \end{bmatrix}$ are equal to each other, then $y = -1, z = 3,$ and $x = 4$.

Matrix multiplication can be used to rewrite systems of linear equations. For example, the matrix equation that is equivalent to the system of equations

$\begin{cases} 2x+3y=6 \\ x+y=-1 \end{cases}$ is $\begin{bmatrix} 2 & 3 \\ 1 & 1 \end{bmatrix} \begin{bmatrix} x \\ y \end{bmatrix} = \begin{bmatrix} 6 \\ -1 \end{bmatrix},$

The matrix $\begin{bmatrix} 2 & 3 \\ 1 & 1 \end{bmatrix}$, which is the matrix formed by the coefficients of the variables, is called the **coefficient matrix**. The system of equations

$\begin{cases} x-2y=4 \\ 2x-y=2 \end{cases}$ can be written as $\begin{bmatrix} 1 & -2 \\ 2 & -1 \end{bmatrix} \begin{bmatrix} x \\ y \end{bmatrix} = \begin{bmatrix} 4 \\ 2 \end{bmatrix}$

and the coefficient matrix for this system of equations is $\begin{bmatrix} 1 & -2 \\ 2 & -1 \end{bmatrix}.$

## Lesson 9-2 Review

1. Find $\begin{bmatrix} 1 & 2 \\ -3 & 5 \end{bmatrix} \cdot \begin{bmatrix} 1 \\ 3 \end{bmatrix}$

2. Find $\begin{bmatrix} 1 & 2 \\ -3 & 5 \end{bmatrix} \cdot \begin{bmatrix} -1 & 3 \\ 4 & 2 \end{bmatrix}$ and $\begin{bmatrix} -1 & 3 \\ 4 & 2 \end{bmatrix} \cdot \begin{bmatrix} 1 & 2 \\ -3 & 5 \end{bmatrix}$

3. Write the system of equations $\begin{cases} 3x - y = 3 \\ 2x + 3y = -2 \end{cases}$ as a matrix equation.

# Lesson 9-3: Determinants

One popular application of matrices involves solving systems of linear equations. In order to use matrices to solve systems of equations, we need to make develop the idea of a determinant.

Square matrices are matrices that have the same number of rows as columns. All square matrices have a determinant. The determinant of a matrix is a scalar, or a number. It can reveal a lot of information about the matrix. The determinant of a matrix $A$ is denoted $|A|$ or **det** $(A)$. The determinant of a matrix can be positive, negative, or zero. Determinants of matrices are discussed in detail in a course on linear algebra. Calculating the determinant of an $n \times n$ matrix using the definition of a determinant is computationally intensive, and I will only discuss how to find the determinant of a $2 \times 2$ matrix.

The determinant of the matrix $\begin{bmatrix} a & b \\ c & d \end{bmatrix}$ is the quantity $ad - bc$. We will practice computing the determinant of a few matrices.

## Example 1

Calculate the determinant of the following matrices:

a. $A = \begin{bmatrix} 2 & 1 \\ -3 & 5 \end{bmatrix}$    b. $B = \begin{bmatrix} 2 & 3 \\ 4 & 6 \end{bmatrix}$    c. $C = \begin{bmatrix} 1 & 1 \\ 7 & 5 \end{bmatrix}$

**Solution:**

a. **det** $(A) = (2)(5) - (1)(-3) = 10 + 3 = 13$

b. **det** $(B) = (2)(6) - (4)(3) = 12 - 12 = 0$

c. **det** $(C) = (1)(5) - (1)(7) = 5 - 7 = -2$

## Lesson 9-3 Review

Find the determinants of the following matrices:

1. $A = \begin{bmatrix} 3 & 4 \\ -2 & 2 \end{bmatrix}$

2. $B = \begin{bmatrix} 1 & 3 \\ -2 & -6 \end{bmatrix}$

3. $C = \begin{bmatrix} 4 & 2 \\ 3 & -5 \end{bmatrix}$

# Lesson 9-4: The Inverse of a Matrix

We have talked about adding, subtracting, and multiplying matrices. We introduced the *zero matrix* (also called the *additive identity matrix*) and the *additive inverse of a matrix*. The idea of an identity and an inverse go hand in hand. In the process of constructing the real numbers, start with the counting numbers, and then introduce 0 (the additive identity) into the mix. Next, throw in the negative integers (which are the additive inverses of the positive numbers), with the understanding that adding a number and its additive inverse give 0 (the additive identity). We followed a similar process with matrices: We have the additive identity and the additive inverse taken care of. Now we will focus on multiplication and multiplicative inverses.

The **multiplicative identity matrix** is a square matrix whose diagonal elements are 1 and all other elements are 0. The $2 \times 2$ identity matrix is the matrix $\begin{bmatrix} 1 & 0 \\ 0 & 1 \end{bmatrix}$. Multiplying any $2 \times 2$ matrix by the matrix $\begin{bmatrix} 1 & 0 \\ 0 & 1 \end{bmatrix}$ will not change the original matrix. For example, you should verify the products

$$\begin{bmatrix} 3 & 4 \\ -2 & 2 \end{bmatrix} \cdot \begin{bmatrix} 1 & 0 \\ 0 & 1 \end{bmatrix} = \begin{bmatrix} 3 & 4 \\ -2 & 2 \end{bmatrix}$$

$$\begin{bmatrix} 1 & 0 \\ 0 & 1 \end{bmatrix} \cdot \begin{bmatrix} 3 & 4 \\ -2 & 2 \end{bmatrix} = \begin{bmatrix} 3 & 4 \\ -2 & 2 \end{bmatrix}$$

We will let I denote the identity matrix.

The next idea to explore is the notion of a multiplicative inverse. Given a $2 \times 2$ matrix $A$, is there a $2 \times 2$ matrix $B$ that, when multiplied by A, gives the identity matrix? In other words, given a matrix $A$, is there a matrix $B$ such that $A \cdot B = I$ and $B \cdot A = I$? The answer to that question is...sometimes.

Keep in mind that a multiplicative inverse is along the same line as a reciprocal. Remember that there is one real number that does not have a multiplicative inverse, or reciprocal: that one real number is 0. Finding the reciprocal of 0 is like dividing by 0, which we know we are not allowed to do. The only square matrices that do *not* have multiplicative inverses are the matrices whose *determinant* is 0. So the determinant of a square matrix indicates whether or not the matrix has an inverse.

### Example 1

Which of the following matrices are invertible?

a. $A = \begin{bmatrix} 2 & 1 \\ -3 & 5 \end{bmatrix}$
   b. $B = \begin{bmatrix} 2 & 3 \\ 4 & 6 \end{bmatrix}$
   c. $C = \begin{bmatrix} 1 & 1 \\ 7 & 5 \end{bmatrix}$

**Solution:** These matrices may look familiar. We found the determinants of these matrices in the previous lesson. The determinants of the matrices $A$ and $C$ were not 0, so those matrices are invertible. The determinant of $B$ is 0, so it is not invertible.

In general, finding the inverse of an $n \times n$ matrix is computationally intensive. You've heard me mention computational difficulties with matrices several times. As I mentioned earlier, matrices are used to solve many industrial problems, and the matrices involved can have thousands of rows and columns. There are many challenges that arise when trying to find solutions to these problems. An entire area of mathematics is devoted to exploring these challenges and devising methods to help simplify calculations involving large matrices. Fortunately, finding the inverse of a $2 \times 2$ invertible matrix is a relatively straightforward process. As a warning, you shouldn't try to find the inverse of a *non-invertible* $2 \times 2$ matrix!

The first step in finding the inverse of a $2 \times 2$ matrix is to find its determinant. This helps determine whether or not the matrix actually has an inverse. Once you know the determinant of the matrix, you can easily find its inverse. The inverse of the matrix $A = \begin{bmatrix} a & b \\ c & d \end{bmatrix}$ is denoted $A^{-1}$, and it is given by the equation $A^{-1} = \frac{1}{\det(A)} \begin{bmatrix} d & -b \\ -c & a \end{bmatrix}$. Keep in mind that if you multiply $A$ and $A^{-1}$ together, you must get the $2 \times 2$ identity matrix $\begin{bmatrix} 1 & 0 \\ 0 & 1 \end{bmatrix}$.

Notice the role that **det** $(A)$ plays in the formula for the inverse: if **det** $(A) = 0$, finding the inverse of the matrix involves dividing by 0, which we know we are not allowed to do. So, knowing the role that the determinant of a matrix has in finding the inverse of the matrix can help you when you have to find the inverse of a matrix.

## Example 2

Find the inverse of the following matrices:

a. $A = \begin{bmatrix} 2 & 1 \\ -3 & 5 \end{bmatrix}$
b. $B = \begin{bmatrix} 2 & 3 \\ 4 & 6 \end{bmatrix}$
c. $C = \begin{bmatrix} 1 & 1 \\ 7 & 5 \end{bmatrix}$

**Solution:** Use the formula for the inverse of a matrix:

a. $A^{-1} = \dfrac{1}{13} \begin{bmatrix} 5 & -1 \\ 3 & 2 \end{bmatrix}$

b. B is not invertible, because $\det(B) = 0$

c. $C^{-1} = -\dfrac{1}{2} \begin{bmatrix} 5 & -1 \\ -7 & 1 \end{bmatrix}$

You should verify that $A \cdot A^{-1} = I$ and $C \cdot C^{-1} = I$. Not only will it give you valuable practice multiplying two matrices, but it will also give you some insight into why this method works. I'll put in one more plug for linear algebra. If you want to know more about matrices, determinants, and inverses, the course to take is linear algebra.

## Lesson 9-4 Review

Find the inverses of the following matrices:

1. $A = \begin{bmatrix} 3 & 4 \\ -2 & 2 \end{bmatrix}$
2. $B = \begin{bmatrix} 1 & 3 \\ -2 & -6 \end{bmatrix}$
3. $C = \begin{bmatrix} 4 & 2 \\ 3 & -5 \end{bmatrix}$

# Lesson 9-5: Solving Systems of Equations

In Lesson 9-2, we learned how to write a system of equations in terms of a matrix equation. The system of equations $\begin{cases} 2x - y = 3 \\ x + 2y = 4 \end{cases}$ can be written as $\begin{bmatrix} 2 & -1 \\ 1 & 2 \end{bmatrix}\begin{bmatrix} x \\ y \end{bmatrix} = \begin{bmatrix} 3 \\ 4 \end{bmatrix}$. If we write $A = \begin{bmatrix} 2 & -1 \\ 1 & 2 \end{bmatrix}$, $X = \begin{bmatrix} x \\ y \end{bmatrix}$ and $C = \begin{bmatrix} 3 \\ 4 \end{bmatrix}$, then this matrix equation looks like the linear equation $AX = C$. We can solve the equation $AX = C$ by multiplying both sides of the equation by the reciprocal, or the inverse, of $A$: $A^{-1}(AX) = A^{-1}C$. As a result, we see that $X = A^{-1}C$. Remember that matrix multiplication does not commute, so the order in which you multiply matters. In order to solve the equation $AX = C$ for $X$, we must consistently multiply both sides of the equation by $A^{-1}$ *on the left.*

## Example 1

Use matrices to solve the system of equations $\begin{cases} 2x - y = 3 \\ x + 2y = 4 \end{cases}$.

**Solution:** Write the system of equations in terms of matrices:

$\begin{bmatrix} 2 & -1 \\ 1 & 2 \end{bmatrix}\begin{bmatrix} x \\ y \end{bmatrix} = \begin{bmatrix} 3 \\ 4 \end{bmatrix}$. Next, find the inverse of the coefficient

matrix. If $A = \begin{bmatrix} 2 & -1 \\ 1 & 2 \end{bmatrix}$, then $A^{-1} = \frac{1}{\det(A)}\begin{bmatrix} 2 & 1 \\ -1 & 2 \end{bmatrix} = \frac{1}{5}\begin{bmatrix} 2 & 1 \\ -1 & 2 \end{bmatrix}$.

Multiply both sides of the equation $\begin{bmatrix} 2 & -1 \\ 1 & 2 \end{bmatrix}\begin{bmatrix} x \\ y \end{bmatrix} = \begin{bmatrix} 3 \\ 4 \end{bmatrix}$ on the left by $A^{-1}$:

$$\frac{1}{5}\begin{bmatrix} 2 & 1 \\ -1 & 2 \end{bmatrix}\begin{bmatrix} 2 & -1 \\ 1 & 2 \end{bmatrix}\begin{bmatrix} x \\ y \end{bmatrix} = \frac{1}{5}\begin{bmatrix} 2 & 1 \\ -1 & 2 \end{bmatrix}\begin{bmatrix} 3 \\ 4 \end{bmatrix}.$$

The multiplication of the matrices on the left should work out to give the identity matrix (remember, we are multiplying by the inverse of $A$...the product of a matrix and its inverse must be the identity matrix).

$$\begin{bmatrix} x \\ y \end{bmatrix} = \frac{1}{5}\begin{bmatrix} 2 & 1 \\ -1 & 2 \end{bmatrix}\begin{bmatrix} 3 \\ 4 \end{bmatrix} = \frac{1}{5}\begin{bmatrix} 10 \\ 5 \end{bmatrix} = \begin{bmatrix} 2 \\ 1 \end{bmatrix}$$

Remember that with scalar multiplication, each element in the matrix is multiplied by the scalar in front of the matrix. The solution to this system of equations is (2, 1).

Using matrices to solve systems of equations requires the coefficient matrix to be invertible. If the coefficient matrix is not invertible (so that the determinant of the coefficient matrix is 0), then there is not a unique solution to the system of equations. Either the system is inconsistent (meaning that the two equations represent parallel lines so that there is no intersection point) or the system is dependent (meaning that the two equations actually represent the same line, so there are infinitely many solutions). Further investigation of the system of equations is necessary to determine which option is correct.

### Example 2

Use matrices to solve the system of equations $\begin{cases} x - y = 1 \\ x + y = 9 \end{cases}$

**Solution:** Write the system of equations in terms of matrices:

$\begin{bmatrix} 1 & -1 \\ 1 & 1 \end{bmatrix} \begin{bmatrix} x \\ y \end{bmatrix} = \begin{bmatrix} 1 \\ 9 \end{bmatrix}$. Next, find the inverse of the coefficient matrix.

If $A = \begin{bmatrix} 1 & -1 \\ 1 & 1 \end{bmatrix}$, then $A^{-1} = \dfrac{1}{\det(A)} \begin{bmatrix} 1 & -1 \\ 1 & 1 \end{bmatrix} = \dfrac{1}{2} \begin{bmatrix} 1 & 1 \\ -1 & 1 \end{bmatrix}$.

Multiply both sides of the equation $\begin{bmatrix} 1 & -1 \\ 1 & 1 \end{bmatrix} \begin{bmatrix} x \\ y \end{bmatrix} = \begin{bmatrix} 1 \\ 9 \end{bmatrix}$ on the left by $A^{-1}$:

$$\frac{1}{2} \begin{bmatrix} 1 & 1 \\ -1 & 1 \end{bmatrix} \begin{bmatrix} 1 & -1 \\ 1 & 1 \end{bmatrix} \begin{bmatrix} x \\ y \end{bmatrix} = \frac{1}{2} \begin{bmatrix} 1 & 1 \\ -1 & 1 \end{bmatrix} \begin{bmatrix} 1 \\ 9 \end{bmatrix}.$$

The multiplication of the matrices on the left should work out to give the identity matrix (remember, we are multiplying by the inverse of $A$...the product of a matrix and its inverse must be the identity matrix).

$$\begin{bmatrix} x \\ y \end{bmatrix} = \frac{1}{2} \begin{bmatrix} 1 & 1 \\ -1 & 1 \end{bmatrix} \begin{bmatrix} 1 \\ 9 \end{bmatrix} = \frac{1}{2} \begin{bmatrix} 10 \\ 8 \end{bmatrix} = \begin{bmatrix} 5 \\ 4 \end{bmatrix}.$$

The solution to this system of equations is (5, 4).

Keep in mind that solving systems of two equations and two unknowns can be done easily using the substitution or elimination methods. However, those methods can become cumbersome when trying to solve large systems of equations. Writing the system of equations in matrix form provides an alternative approach to solving systems of equations. Unfortunately, finding the inverse of a matrix can also be difficult. Fortunately, there are many techniques available to simplify the process of finding the inverse of a matrix, which is why this method is so useful.

## Lesson 9-5 Review

Solve the following systems of equations using matrices:

1. $\begin{cases} 3x - 2y = 1 \\ 4x + 5y = 9 \end{cases}$
2. $\begin{cases} 2x - y = 1 \\ 3x - 2y = 0 \end{cases}$

# Lesson 9-6: Cramer's Rule

Cramer's Rule for solving systems of equations involves using the determinant of matrices rather than the inverse of a matrix. To use Cramer's Rule to solve the system of equations $\begin{cases} a_1 x + b_1 y = c_1 \\ a_2 x + b_2 y = c_2 \end{cases}$, find the determinants of the following three matrices:

▶ The coefficient matrix $\begin{bmatrix} a_1 & b_1 \\ a_2 & b_2 \end{bmatrix}$.

If the determinant of this coefficient matrix is 0, then this matrix is not invertible, and Cramer's method will not work. Let $D$ denote the determinant of the coefficient matrix.

▶ The matrix $\begin{bmatrix} c_1 & b_1 \\ c_2 & b_2 \end{bmatrix}$.

The coefficients of $x$ in the coefficient matrix are replaced with the constants. Let $D_x$ represent the determinant of this matrix.

▶ The matrix $\begin{bmatrix} a_1 & c_1 \\ a_2 & c_2 \end{bmatrix}$.

The coefficients of $y$ in the coefficient matrix are replaced with the constants. Let $D_y$ represent the determinant of this matrix.

The value of $x$ is given by the equation $x = \frac{D_x}{D}$, and the value of $y$ is given by the equation $y = \frac{D_y}{D}$. Cramer's Rule allows us to solve for $x$ and $y$ by computing only these three determinants.

## Example 1

Use Cramer's Rule to solve the system of equations $\begin{cases} x - y = 1 \\ x + y = 9 \end{cases}$.

**Solution:** The coefficient matrix is $\begin{bmatrix} 1 & -1 \\ 1 & 1 \end{bmatrix}$, and its determinant is 2.

The second matrix involves replacing the coefficients of $x$ with the constant terms in the system of equations: $\begin{bmatrix} 1 & -1 \\ 9 & 1 \end{bmatrix}$.

The determinant of this matrix is 10. We can now find $x$:

$$x = \frac{10}{2} = 5.$$

The third matrix involves replacing the coefficients of $y$ with the constant terms in the system of equations: $\begin{bmatrix} 1 & 1 \\ 1 & 9 \end{bmatrix}$.

The determinant of this matrix is 8. We can now find $y$:

$$y = \frac{8}{2} = 4.$$ The solution to this system of equations is (5, 4).

## Example 2

Use Cramer's Rule to solve the system of equations

$$\begin{cases} x + y = 1 \\ x + 2y = 4 \end{cases}.$$

**Solution:** The coefficient matrix is $\begin{bmatrix} 1 & 1 \\ 1 & 2 \end{bmatrix}$, and its determinant is 1.

The second matrix involves replacing the coefficients of $x$ with the constant terms in the system of equations: $\begin{bmatrix} 1 & 1 \\ 4 & 2 \end{bmatrix}$.

The determinant of this matrix is −2. We can now find $x$:

$$x = \frac{-2}{1} = -2 \, .$$

The third matrix involves replacing the coefficients of $y$ with the

constant terms in the system of equations: $\begin{bmatrix} 1 & 1 \\ 1 & 4 \end{bmatrix}$.

The determinant of this matrix is 3. We can now find $y$: $y = \frac{3}{1} = 3$.

The solution to this system of equations is (−2, 3).

---

Cramer's Rule, like all problem-solving techniques, has its advantages and disadvantages. Finding the determinant of a large matrix can be difficult, but there are some ways to simplify the process.

## Lesson 9-6 Review

Solve the following systems of equations using Cramer's Rule:

1. $\begin{cases} 2x + y = 9 \\ 2x - 3y = 5 \end{cases}$

2. $\begin{cases} x + 5y = 3 \\ 2x - 3y = 6 \end{cases}$

## Answer Key

### Lesson 9-1 Review

1.  a. $A + B = \begin{bmatrix} 10 & -1 \\ 0 & 12 \end{bmatrix}$

    c. $5C = \begin{bmatrix} 10 & 25 & 0 \\ -45 & 15 & -5 \end{bmatrix}$

    b. $3A - B = \begin{bmatrix} -2 & -11 \\ 4 & 4 \end{bmatrix}$

    d. The additive inverse of $C$ is $\begin{bmatrix} -2 & -5 & 0 \\ 9 & -3 & 1 \end{bmatrix}$

### Lesson 9-2 Review

1. $\begin{bmatrix} 1 & 2 \\ -3 & 5 \end{bmatrix} \begin{bmatrix} 1 \\ 3 \end{bmatrix} = \begin{bmatrix} 7 \\ 12 \end{bmatrix}$

2. $\begin{bmatrix} 1 & 2 \\ -3 & 5 \end{bmatrix} \begin{bmatrix} -1 & 3 \\ 4 & 2 \end{bmatrix} = \begin{bmatrix} 7 & 7 \\ 23 & 1 \end{bmatrix}$ and $\begin{bmatrix} -1 & 3 \\ 4 & 2 \end{bmatrix} \begin{bmatrix} 1 & 2 \\ -3 & 5 \end{bmatrix} = \begin{bmatrix} -10 & 13 \\ -2 & 18 \end{bmatrix}$

3. $\begin{bmatrix} 3 & -1 \\ 2 & 3 \end{bmatrix} \begin{bmatrix} x \\ y \end{bmatrix} = \begin{bmatrix} 3 \\ -2 \end{bmatrix}$

## Lesson 9-3 Review

1.  $\det(A) = 14$        2.  $\det(B) = 0$        3.  $\det(C) = -26$

## Lesson 9-4 Review

1.  $A^{-1} = \frac{1}{14}\begin{bmatrix} 2 & -4 \\ 2 & 3 \end{bmatrix}$

2.  $B^{-1}$ does not exist because the determinant of $B$ is 0.

3.  $C^{-1} = -\frac{1}{26}\begin{bmatrix} -2 & -2 \\ -3 & 4 \end{bmatrix}$

## Lesson 9-5 Review

1.  $(1,1)$              2.  $(2,3)$

## Lesson 9-6 Review

1.  $D = -8, D_x = -32, D_y = -8$; the solution is $(4,1)$
2.  $D = -13, D_x = -39, D_y = 0$; the solution is $(3,0)$

# Analytic Geometry

The Cartesian coordinate system enables mathematicians to solve geometry problems algebraically. With this coordinate system, we are also able to take geometric concepts and describe them using algebraic equations. For example, from a geometric perspective, a circle is the set of all points that are a fixed distance $r$ from a given point $C$, called the center of the circle. If the coordinates of the circle are $(h, k)$, then the formula for the circle is $(x - h)^2 + (y - k)^2 = r^2$. There are other geometric concepts that can be described using algebraic equations. In this chapter, we will use the geometric descriptions of parabolas, ellipses, and hyperbolas to discover their algebraic equivalents.

The geometric definitions of a circle, a parabola, an ellipse, and a hyperbola are similar in that they all involve the distance between a specific point, called a **focus** (or points, called **foci**) and some other object. Circles, parabolas, ellipses, and hyperbolas are called conics, or conic sections, because their shapes can be generated by intersecting a plane and a cone, as shown in Figure 10.1.

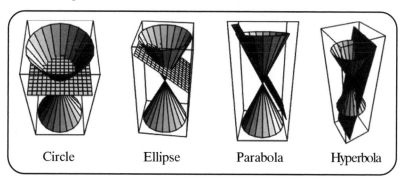

Circle          Ellipse          Parabola          Hyperbola

*Figure 10.1*

These four curves all result from slicing a pair of cones with a plane. If a plane slices one of the cones parallel to its base, the resulting conic is a circle. If the cones are sliced at an angle, the result is an elongated circle, or an ellipse. The more we tilt the slice, the more elongated the ellipse becomes. When the slice is parallel to the side of the cone, the curve is no longer closed, and the result is a parabola. If the angle of the slice is tilted further, the plane intersects both cones and a hyperbola is created.

Each conic will have its own terminology. There are several important features of the various conics that will enable us to sketch their graphs easily. In this chapter, we will examine parabolas, ellipses, and hyperbolas.

# Lesson 10-1: Parabolas

As we discussed in Chapter 3, we know that the shape of the graph of a quadratic equation is called a parabola. There are five key features of a parabola: concavity, axis of symmetry, vertex, $y$-intercept, and $x$-intercepts. The parabolas that we analyzed in Chapter 3 opened either upward or downward. Parabolas can also open sideways, but when parabolas are oriented this way they fail the vertical line test, so they are not functions.

Geometrically, a parabola is defined as the collection of all points $P$ in the plane that are the same distance from a fixed point $F$ as they are from a fixed line, $D$. The point $F$ is called the **focus** of the parabola, and the line $D$ is called the **directrix**.

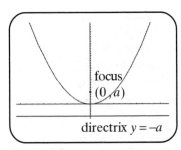

Figure 10.2 shows a parabola, its focus, and its directrix. The line that passes through the focus $F$ and is perpendicular to the directrix $D$ is called the **axis of symmetry** of the parabola. The point where the axis of symmetry and the parabola intersect is called the **vertex** of the parabola.

*Figure 10.2*

Any point in the Cartesian coordinate system can serve as the focus of a parabola, and any line in the Cartesian coordinate system can serve as the directrix. We will use the distance formula to derive a formula for a parabola. In this lesson, we will consider the special case in which the directrix is parallel to one of the coordinate axes.

Suppose that the vertex $V$ is located at the origin, and that the focus $F$ is located on the positive $x$-axis, at the point $(a, 0)$, where $a > 0$, as shown in Figure 10.3. Then the directrix must be the line $x = -a$.

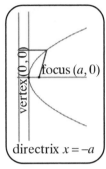

*Figure 10.3*

If $P = (x, y)$ is any point on the parabola, then the distance between $P$ and the focus must equal the distance between $P$ and the directrix.

Now, the distance between $P$ and the focus is $\sqrt{(x-a)^2 + (y-0)^2}$, or $\sqrt{(x-a)^2 + y^2}$. The distance between $P$ and the directrix is $|x + a|$. We can set these two distances equal to each other to find the equation of this parabola:

$$\sqrt{(x-a)^2 + y^2} = |x + a|$$

Square both sides of this equation $\qquad (x - a)^2 + y^2 = (x + a)^2$

Expand $(x - a)^2$ and $(x + a)^2$ $\qquad x^2 - 2ax + a^2 + y^2 = x^2 + 2ax + a^2$

Simplify $\qquad\qquad\qquad\qquad\qquad y^2 = 4ax$

The graph of the parabola $y^2 = 4ax$ is shown in Figure 10.4. The parabola lies entirely in Quadrant I and Quadrant IV; the $x$-coordinate of every point on the parabola must be positive, because $a > 0$ and $y^2 > 0$.

If we keep the vertex of the parabola at the origin, we can also place the focus $F$ on the positive $y$-axis, the negative $x$-axis, or the negative $y$-axis. The characteristics of the general form of a parabola are shown in the table here, and their graphs are shown in Figure 10.5 on page 202.

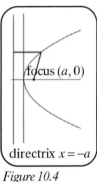

*Figure 10.4*

| Equation | $y^2 = ax$ | $x^2 = ay$ |
|---|---|---|
| Focus | $\left(\frac{a}{4}, 0\right)$ | $\left(0, \frac{a}{4}\right)$ |
| Directrix | $x = -\frac{a}{4}$ | $y = -\frac{a}{4}$ |
| Axis of Symmetry | $x$-axis | $y$-axis |
| Opens | Right if $a > 0$; Left if $a < 0$ | Up if $a > 0$; Down if $a < 0$ |
| Quadrants | I & IV if $a > 0$, II & III if $a < 0$ | I & II if $a > 0$, III & IV if $a < 0$ |
| Example | $y^2 = x$ | $x^2 = y$ |

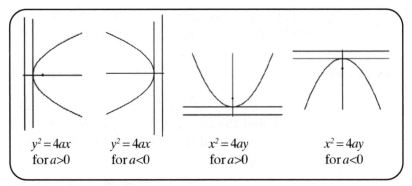

| $y^2 = 4ax$ for $a>0$ | $y^2 = 4ax$ for $a<0$ | $x^2 = 4ay$ for $a>0$ | $x^2 = 4ay$ for $a<0$ |

*Figure 10.5*

## Example 1

For the following equations, determine the focus, the directrix, the axis of symmetry, and whether the parabola opens to the left, the right, up or down:

a. $y^2 = 12x$                 b. $x^2 = -6y$

**Solution:**

a. $y^2 = 12x$: This parabola is of the form $y^2 = ax$, where $a = 12$. The focus is $\left(\frac{a}{4}, 0\right)$, or $(3, 0)$. The directrix is $x = -3$, the axis of symmetry is the $x$-axis, and it opens to the right. The graph of $y^2 = 12x$ is shown in Figure 10.6.

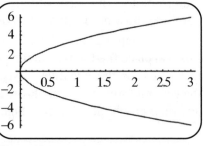

*Figure 10.6*

b. $x^2 = -6y$: This parabola is of the form $x^2 = ay$, where $a = -6$. The focus is $\left(0, -\frac{6}{4}\right)$, or $\left(0, -\frac{3}{2}\right)$, the directrix is $y = \frac{3}{2}$, the axis of symmetry is the $y$-axis, and it opens down. The graph of $x^2 = -6y$ is shown in Figure 10.7.

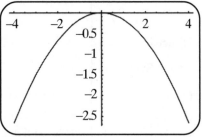

*Figure 10.7*

## Example 2

Find the equation of a parabola whose vertex is the point $(0, 0)$, axis of symmetry is the $x$-axis, and its graph contains the point $(2, 3)$.

**Solution:** The axis of symmetry is the $x$-axis, so the general form of the parabola is $y^2 = ax$. We can substitute the point $(2, 3)$ into the equation of the parabola to solve for $a$:

$$y^2 = ax$$
$$3^2 = a \cdot 2$$
$$a = \frac{9}{2}$$

The equation of the parabola is $y^2 = \frac{9}{2}x$.

There is no requirement that the vertex of a parabola has to be located at the origin; a parabola can be translated horizontally or vertically. When a parabola is translated, every point on the parabola is translated by the same amount. In particular, the vertex of the parabola is translated. The equations $y^2 = ax$ and $x^2 = ay$ describe parabolas whose vertex is $(0, 0)$. We could write these equations as $(y - 0)^2 = a(x - 0)$ and $(x - 0)^2 = a(y - 0)$, if we wanted to explicitly include the location of the vertex of the parabola. Remember that shifting a function vertically involves adding a constant to $y$, and shifting a function horizontally involves adding a constant to $x$. In general, the equation of a parabola whose vertex is the point $(h, k)$ can be written in one of the following two forms depending on whether the axis of symmetry is parallel to the $x$-axis or the $y$-axis: $(y - k)^2 = a(x - h)$ or $(x - h)^2 = a(y - k)$. It may be helpful to realize that the vertex and the focus of a parabola must lie on the axis of symmetry, regardless of how the parabola is oriented in the plane. The following table describes the four different parabolas whose vertex is the point $(h, k)$.

| Equation | $(y-k)^2 = a(x-h)$ | $(x-h)^2 = a(y-k)$ |
|---|---|---|
| Vertex | $(h, k)$ | $(h, k)$ |
| Focus | $\left(h + \frac{a}{4}, k\right)$ | $\left(h, k + \frac{a}{4}\right)$ |
| Directrix | $x = h - \frac{a}{4}$ | $y = k - \frac{a}{4}$ |
| Axis of Symmetry | $y = k$ | $x = h$ |
| Opens | Right if $a > 0$; Left if $a < 0$ | Up if $a > 0$; Down if $a < 0$ |
| Example | $(y-2)^2 = 4(x-3)$ | $(x-2)^2 = 4(y-3)$ |

Notice that the distance between the vertex and the focus is $\left|\frac{a}{4}\right|$. This relationship may be used to find $|a|$.

## Example 3

Find the equation of the parabola whose vertex is located at the point (−2, 3) and whose focus is the point (0, 3).

**Solution:** Think of yourself as a detective when solving these problems. The location of the vertex and the focus are clues to a puzzle. Use the preceding table to help you put the pieces together. To begin with, the vertex and the focus must lie on the axis of symmetry. The equation of the line that passes through the points (−2, 3) and (0, 3) is $y = 3$. So the axis of symmetry is a horizontal line, meaning that the directrix is a vertical line. The distance between the vertex and the focus is 2, so $\left|\frac{a}{4}\right| = 2$, and $|a| = 8$. The focus lies to the right of the vertex, so the parabola must open to the right, meaning that $a > 0$. The equation of this parabola must be of the form $(y - k)^2 = a(x - h)$, and we know $a$, $h$, and $k$:

$(y - 3)^2 = 8(x - (-2))$, or $(y - 3)^2 = 8(x + 2)$.

## Example 4

Find the location of the vertex, the focus, and the directrix of the parabola $x^2 + 4x + 4y = 0$.

**Solution:** We need to write this equation in the form $(x - h)^2 = a(y - k)$. In order to do this, we need to complete the square for the terms that involve $x$:

$$x^2 + 4x + 4y = 0$$

Divide the linear coefficient of $x$ by 2 and square the result. Add this number to both sides of the equation.     $x^2 + 4x + 4 + 4y = 0 + 4$

Factor the quadratic expression involving $x$.  $(x + 2)^2 + 4y = 4$

Subtract $4y$ from both sides.                 $(x + 2)^2 = -4y + 4$

Factor −4 from both terms on the right.        $(x + 2)^2 = -4(y - 1)$

The vertex is the point (−2, 1). From the equation of the parabola, we see that $h = -2$, $k = 1$, and $a = -4$. To find the focus, substitute the values for $h$, $k$, and $a$ into the formula

$\left(h, k + \dfrac{a}{4}\right)$ and simplify:

$\left(-2, 1 + \left(\dfrac{-4}{4}\right)\right)$

$(-2, 0)$

To find the directrix, substitute the values for $h$, $k$, and $a$ into

the formula $y = k - \dfrac{a}{4}$ and simplify:

$y = 1 - \left(\dfrac{-4}{4}\right) = 2$.

The vertex is the point $(-2, 1)$, the focus is the point $(-2, 0)$, and the directrix is the line $y = 2$.

---

Objects that are shaped like a parabola have the ability to collect low intensity signals from sources that are very far away. Light or sound that hits a parabolic surface is reflected, or channeled, to the focus. The intensity of the signal is much stronger at the focus than it is anywhere on the surface of the parabola. Satellite dishes, searchlights, and automobile headlights are a few examples of objects that incorporate the properties of a parabola into their design. Science museums often have two parabolic dishes set up on opposite sides of a large room, and a metal circle located at the focus of each of the dishes. If one person whispers near one focus, the sound will be audible to someone listening at the other focus.

## Example 5

A satellite dish is shaped like a paraboloid of revolution (a surface obtained by revolving a parabola around its axis). The signals from a satellite will strike the surface of the dish and be collected at the receiver, which is located at the focus of the paraboloid. If the dish is 12 feet across at its opening and 6 feet deep at its center, where should the receiver be placed?

**Solution:** In this problem, it does not matter which way the satellite dish is oriented. It also does not matter where the satellite dish is located in the coordinate plane. The only thing that matters is the relative distances for the dimensions of the satellite dish. We can orient the satellite dish so that it points upward, with the vertex of the dish at the origin and the focus located along the positive $y$-axis, as shown in Figure 10.8 on page 206. The dish is 12 feet across and 6 feet deep, which means that the parabola will pass through the points $(6, 6)$ and $(-6, 6)$. With this orientation, the equation of the parabola will be $x^2 = ay$, and its focus will be the point $\left(0, \dfrac{a}{4}\right)$. Be-

cause (6, 6) is a point on the parabola, we can substitute this information into the equation for the parabola and solve for $a$:

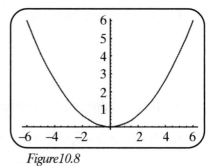

Figure 10.8

$x^2 = ay$

$6^2 = a \cdot 6$

$a = \dfrac{36}{6} = 6$

The focus is the point $\left(0, \dfrac{a}{4}\right)$, or $\left(0, \dfrac{3}{2}\right)$. The receiver should be placed 1.5 feet from the base of the dish, along its axis of symmetry.

---

## Lesson 10-1 Review

1. For the following equations, determine the focus, the directrix, the axis of symmetry, and whether the parabola opens to the left, the right, up, or down:

   a. $y^2 = -10x$                 b. $x^2 = 4y$

2. Find the equation of a parabola whose vertex is the point (0, 0), axis of symmetry is the x-axis, and its graph contains the point (4, 3).

3. Find the equation of the parabola whose vertex is located at the point (3, –2) and whose focus is the point (3, 4).

4. Find the location of the vertex, the focus, and the directrix of the parabola $y^2 + 6y + x = 0$.

5. A satellite dish is shaped like a paraboloid of revolution. The signals from a satellite will strike the surface of the dish and be collected at the receiver, which is located at the focus of the paraboloid. If the dish is 10 feet across at its opening and 4 feet deep at its center, where should the receiver be placed?

# Lesson 10-2: Ellipses

An ellipse looks like an elongated, or stretched circle. Orbits of planets, moons, and satellites follow an elliptical path. An **ellipse** is defined as the collection of all points in the plane, the sum of whose distances from two fixed points $F_1$ and $F_2$ is a constant. The two fixed points are called the **foci** of the ellipse.

From this geometric definition of an ellipse, we can devise a physical way to draw an ellipse. Attach the two endpoints of a string to a piece of cardboard using thumbtacks, keeping some slack in the string. The length of the string is the constant sum referred to in the definition of the ellipse, and the two endpoints represent the foci of the ellipse. Use the point of a pencil to tighten the string and move the pencil around the foci, keeping the string tight at all times. The pencil will trace out an ellipse.

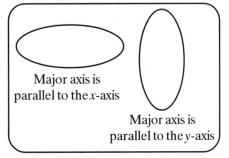

Major axis is parallel to the x-axis

Major axis is parallel to the y-axis

If the string is only slightly longer than the distance between the foci, then the ellipse that is traced out will be elongated in shape. If the foci are close together, relative to the length of the string, then the ellipse will be almost circular.

If $F_1$ and $F_2$ are the foci of an ellipse, then the line containing $F_1$ and

*Figure 10.9*

$F_2$ is called the **major axis**. The midpoint of the line segment $\overline{F_1F_2}$ is called the **center** of the ellipse. The line that passes through the center of the ellipse and is perpendicular to the major axis is called the **minor axis**. The points where the ellipse intersects the major axis are the **vertices** of the ellipse. An ellipse is symmetric with respect to its major axes, its minor axes, and its center.

The general equation of an ellipse whose center is located at the origin is given by the equation:

$$\frac{x^2}{a^2} + \frac{y^2}{b^2} = 1$$

If $0 < b < a$, then the major axis is the x-axis, the vertices are located at $(\pm a, 0)$, the horizontal length of the ellipse is $2a$ and the vertical length of the ellipse is $2b$. The foci are located at $(\pm c, 0)$, where $c$ satisfies the equation $c^2 = a^2 - b^2$.

Most of these details can be determined from the equation $\frac{x^2}{a^2} + \frac{y^2}{b^2} = 1$, or from the graph of the ellipse, such as the one shown in Figure 10.10, and should not be memorized.

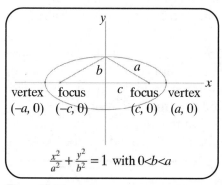

*Figure 10.10*

The points where the ellipse intersects the coordinate axes are just the intercepts of the equation $\frac{x^2}{a^2} + \frac{y^2}{b^2} = 1$. Remember that the $x$-intercepts are found by evaluating the formula when $y = 0$, and the $y$-intercepts are found by evaluating the formula when $x = 0$. It is pretty easy to see that the $x$-intercepts are the solutions to the equation $\frac{x^2}{a^2} = 1$,

and the $y$-intercepts are the solutions to the equation $\frac{y^2}{b^2} = 1$. The location of the foci can be determined using the Pythagorean Theorem. Notice that we can construct a right triangle within the ellipse. The lengths of the two legs of the right triangle are $b$ and $c$, and the length of the hypotenuse is $a$. From the Pythagorean Theorem, we see that $b^2 + c^2 = a^2$. The length of the major axis is the distance between the two vertices of the ellipse.

If, on the other hand, $0 < a < b$, the role of $a$ and the role of $b$ switch. The major axis is the $y$-axis, and the vertices are located at $(0, \pm b)$. The horizontal length of the ellipse is still $2a$ and the vertical length of the ellipse is still $2b$. The foci are located at $(0, \pm c)$, where $c$ satisfies the equation $c^2 = b^2 - a^2$.

Most of these details can be determined from the equation $\frac{x^2}{a^2} + \frac{y^2}{b^2} = 1$, or from the graph of the ellipse, such as the one shown in Figure 10.11, and should not be memorized.

The key to working with ellipses is to remember that the variable with the larger denominator wins. The variable with the

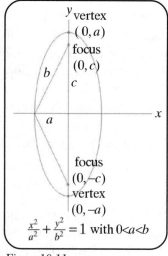

*Figure 10.11*

larger denominator serves as the major axis, and gets to have the foci and the vertices. A little common sense leads you to the equation $c^2 = |a^2 - b^2|$ to find the non-zero coordinates of the foci. If $a > b$ then the foci are on the $x$-axis, and if $a < b$, the foci are on the $y$-axis. To sketch the graph of an ellipse, find the $x$-intercepts and the $y$-intercepts and sketch an ellipse that passes through those points. The longer axis is the major axis, and the smaller axis is the minor axis. The **length of the major axis** is the distance between the two vertices of the ellipse.

The **eccentricity** of an ellipse is defined as the ratio of the distance between the foci of the ellipse to the distance between the two vertices of the ellipse. Remember that the foci and the vertices of an ellipse lie on the major axis. If the foci are located at $(\pm c, 0)$ and the vertices are located at $(\pm a, 0)$, then the eccentricity of the ellipse, $e$, is given by the formula:

$$e = \frac{c}{a}$$

If the foci are located at $(0, \pm c)$ and the vertices are located at $(0, \pm b)$, then the eccentricity of the ellipse, $e$, is given by the formula:

$$e = \frac{c}{b}$$

The eccentricity of an ellipse is a number between 0 and 1, and its value indicates how elongated the ellipse is. The greater the elongation, the more oval the ellipse, or the more that the ellipse deviates from a circle. If $e$ is close to 1, then $c$ is close to $a$, which means that $b$ is small. In this case, the ellipse is elongated in shape. If $e$ is close to 0, then the ellipse is close to a circle in shape.

## Example 1

Sketch the graph of the ellipse given by the

formula $\dfrac{x^2}{9} + \dfrac{y^2}{25} = 1$. Find the vertices, the foci, and the eccentricity of the ellipse.

**Solution:** The denominator of the term that involves $y$ is larger, so the major axis will be the $y$-axis, and the minor axis will be the $x$-axis. From the formula for the ellipse, we have that $a = 3$ and $b = 5$.
The $y$-coordinates of the foci are the solutions to the equation $c^2 = b^2 - a^2 = 5^2 - 3^2 = 25 - 9 = 16$.

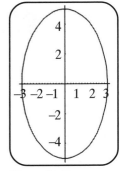

*Figure 10.12*

The foci are located at $(0, \pm 4)$. The vertices are located at $(0, \pm 5)$.

The eccentricity is $e = \dfrac{c}{b} = \dfrac{4}{5}$. The graph of the ellipse is shown in Figure 10.12 on page 209.

## Example 2

The vertices of an ellipse are $(0, \pm 4)$ and the foci are $(0, \pm 2)$. Find the equation of the ellipse and its eccentricity. Sketch a graph of the ellipse.

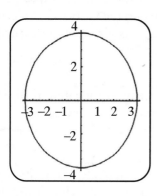

Figure 10.13

**Solution:** The eccentricity is $e = \dfrac{2}{4} = \dfrac{1}{2}$.

The vertices and the foci of the ellipse are along the y-axis, so the major axis is the y-axis. The foci can be found using the equation $c^2 = b^2 - a^2$. We know the values of $b$ and $c$, so we can find $a^2$:

$$c^2 = b^2 - a^2$$
$$2^2 = 4^2 - a^2$$
$$4 = 16 - a^2$$
$$a^2 = 12$$

The equation of the ellipse is $\dfrac{x^2}{12} + \dfrac{y^2}{16} = 1$.

The sketch of the ellipse is shown in Figure 10.13.

## Example 3

The foci of an ellipse are $(0, \pm 8)$, and the eccentricity of the ellipse is $e = \dfrac{4}{5}$. Find the equation of the ellipse and its vertices. Sketch a graph of the ellipse.

**Solution:** The foci of the ellipse lie on the y-axis, so the y-axis is the major axis. We can use the foci of the ellipse and the eccentricity to find the non-zero coordinate of one of the vertices:

$$e = \frac{c}{b}$$

$$\frac{4}{5} = \frac{8}{b}$$

$$b = 10$$

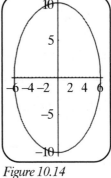

Figure 10.14

The vertices of the ellipse are $(0, \pm10)$. We can use the equation $c^2 = b^2 - a^2$ to find $a^2$:

$$c^2 = b^2 - a^2$$
$$8^2 = 10^2 - a^2$$
$$64 = 100 - a^2$$
$$a^2 = 36$$

The equation of the ellipse is $\frac{x^2}{36} + \frac{y^2}{100} = 1$. The graph of the ellipse is shown in Figure 10.14.

An ellipse can be translated horizontally or vertically. When an ellipse is translated, every point on the ellipse is translated by the same amount. In particular, the center, the foci, and the vertices of the ellipse are translated. The equation $\frac{x^2}{a^2} + \frac{y^2}{b^2} = 1$ describes an ellipse whose center is $(0, 0)$. We could write this equation as $\frac{(x-0)^2}{a^2} + \frac{(y-0)^2}{b^2} = 1$, if we wanted to explicitly include the location of the center of the ellipse. In general, the equation of an ellipse whose center is the point $(h, k)$ can be written as:

$$\frac{(x-h)^2}{a^2} + \frac{(y-k)^2}{b^2} = 1$$

It may be helpful to realize that the vertices and the foci of an ellipse must lie on the major axis, regardless of how the ellipse is oriented in the plane. The values of $a$, $b$, and $c$ no longer represent the nonzero coordinates of the vertices or the foci. Rather, these values represent the distance between the center and either a vertex or a focus. In particular, the value of $a$ (or $b$, depending on which one is larger) represents the distance between the center of the ellipse and one vertex (where the ellipse intersects the major axis). The value of $c$ represents the distance between the center of the ellipse and one of the foci. The value of $b$ (or $a$, depending on which one is smaller) represents the distance between the center of the ellipse and the point where the ellipse intersects the minor axis. The following table summarizes the information for an ellipse with center $(h, k)$.

| Equation for the ellipse | $\dfrac{(x-h)^2}{a^2}+\dfrac{(y-k)^2}{b^2}=1,$ with $0<b<a$ | $\dfrac{(x-h)^2}{a^2}+\dfrac{(y-k)^2}{b^2}=1$ with $0<a<b$ |
|---|---|---|
| Major axis | Parallel to the $x$-axis | Parallel to the $y$-axis |
| Vertices | $(h\pm a,k)$ | $(h,k\pm b)$ |
| Foci | $(h\pm c,k)$ | $(h,k\pm c)$ |
| Eccentricity | $e=\dfrac{c}{a}$ | $e=\dfrac{c}{b}$ |

## Example 4

Find the equation of the ellipse with one focus at $(4, 8)$ and vertices located at $(4, 3)$ and $(4, 9)$.

**Solution:** The focus and vertices must lie on the same line. In this case, all three points lie on the line $x = 4$. The major axis is parallel to the $y$-axis, so the equation of the ellipse will be

$$\frac{(x-h)^2}{a^2}+\frac{(y-k)^2}{b^2}=1, \text{ where } 0 < a < b.$$

We will need to find the center of the ellipse, which will be the midpoint of the line segment connecting the two vertices. The midpoint of the line segment whose endpoints are $(4, 3)$ and $(4, 9)$ is $(4, 6)$. The equation of the ellipse is beginning to develop. We can incorporate the center of the ellipse into the equation:

$$\frac{(x-4)^2}{a^2}+\frac{(y-6)^2}{b^2}=1.$$

The distance between the center and one of the vertices is $b$. The distance between $(4, 9)$ and $(4, 6)$ is 3, so $b = 3$. The distance between the center and one of the foci is $c$. The distance between $(4, 8)$ and $(4, 6)$ is 2, so $c = 2$. Now that we know $b$ and $c$, we can find $a^2$:

$c^2 = b^2 - a^2$

$2^2 = 3^2 - a^2$

$4 = 9 - a^2$

$a^2 = 5$

The equation of the ellipse is $\dfrac{(x-4)^2}{5} + \dfrac{(y-6)^2}{9} = 1$.

---

Ellipses have a reflection property similar to what we saw with parabolas. If an energy source is placed at one focus of a reflecting surface with elliptical cross sections, then the energy will be reflected off of the surface and directed to the other focus. The energy source can be light or sound. A whispering gallery is a room designed with elliptical ceilings. A whisper spoken at one focus can be heard clearly across the room at the other focus. St. Paul's Cathedral in London, and the Statuary Hall in the United States' Capitol Building are two examples of well-known whispering galleries.

There are other applications of ellipses. The orbits of the planets around the sun are elliptical, with the sun positioned at one of the foci. A tank truck is often in the shape of an ellipse. If its shape were more circular, it would be "top-heavy" and less stable. In addition, any cylinder that is sliced at an angle will yield an elliptical cross-section.

## Lesson 10-2 Review

1. Sketch the graph of the ellipse given by the formula $\dfrac{x^2}{100} + \dfrac{y^2}{64} = 1$.
   Find the vertices, the foci, and the eccentricity of the ellipse.

2. The vertices of an ellipse are $(\pm 5, 0)$ and the foci are $(\pm 2, 0)$. Find the equation of the ellipse and its eccentricity. Sketch a graph of the ellipse.

3. The vertices of an ellipse are $(\pm 12, 0)$, and the eccentricity of the ellipse is $e = \dfrac{2}{3}$. Find the equation of the ellipse and its foci. Sketch a graph of the ellipse.

4. Find the equation of the ellipse with one focus at $(10, 3)$ and vertices located at $(-2, 3)$ and $(12, 3)$.

## Lesson 10-3: Hyperbolas

The graphs of ellipses and hyperbolas look very different, but their definitions are very similar. With an *ellipse*, the distances from two fixed foci are *added*. With a *hyperbola*, the distances from two fixed foci are *subtracted*.

A **hyperbola** is defined to be the set of all points in the plane, the difference of whose distances from two fixed points $F_1$ and $F_2$ is a constant. The two fixed points are called the **foci** of the hyperbola. Figure 10.15 shows a hyperbola with foci $F_1$ and $F_2$. The line that contains the foci is called the **transverse axis**. The midpoint of the line segment $\overline{F_1 F_2}$ is called the **center** of the hyperbola. The line that passes through the center and is perpendicular to the transverse axis is called the **conjugate axis**. A hyperbola consists of two separate curves, or **branches**, that are symmetric with respect to the transverse axis, the conjugate axis, and the center. The points where the hyperbola intersects the transverse axis are called the **vertices** of the hyperbola. Each vertex of a hyperbola lies between the center and one of the foci of the hyperbola.

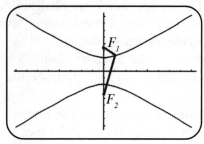

Figure 10.15

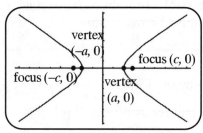

Figure 10.16

We will begin by presenting the equation of a hyperbola when the transverse axis is the *x*-axis. The equation for the hyperbola with center at the origin, foci at $(\pm c, 0)$, and vertices at $(\pm a, 0)$, as shown in Figure 10.16, is:

$$\frac{x^2}{a^2} - \frac{y^2}{b^2} = 1 \text{, where } b^2 = c^2 - a^2$$

If the transverse axis of a hyperbola is the *y*-axis, then the foci and the vertices of the hyperbola will be located on the *y*-axis. The equation for the hyperbola with center at the origin, foci at $(0, \pm c)$, and vertices at $(0, \pm a)$, as shown in Figure 10.17, is:

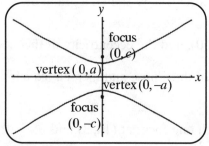

Figure 10.17

$$\frac{y^2}{a^2} - \frac{x^2}{b^2} = 1 \text{, where } b^2 = c^2 - a^2$$

Keep in mind that a hyperbola involves a *difference* of two quadratic terms. The *positive* term determines which axis is the transverse axis. The foci and vertices of a hyperbola lie on the transverse axis.

## Example 1

Find the coordinates of the foci and the vertices of the hyperbola described by $\frac{x^2}{9} - \frac{y^2}{16} = 1$, and sketch its graph.

**Solution:** Because the positive term involves $x$, the $x$-axis is the transverse axis. The foci and the vertices of the hyperbola will lie on the $x$-axis. The vertices of the hyperbola are $(\pm 3, 0)$. The foci are located at $(\pm c, 0)$, where the $x$-coordinates of the foci must

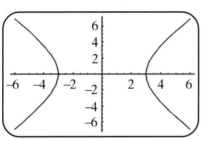

*Figure 10.18*

satisfy the equation $b^2 = c^2 - a^2$:

$$b^2 = c^2 - a^2$$

$$4^2 = c^2 - 3^2$$

$$16 = c^2 - 9$$

$$c^2 = 25$$

The foci are located at $(\pm 5, 0)$. The graph of the hyperbola is shown in Figure 10.18.

## Example 2

Find the equation of the hyperbola that has one vertex at $(0, 2)$ and foci at $(0, \pm 3)$.

**Solution:** The foci and the vertices of the hyperbola are located on the $y$-axis, so the transverse axis is the $y$-axis. The center of the hyperbola is located at the midpoint of the line segment connecting the two foci. The midpoint of the line segment with endpoints $(0, 3)$ and $(0, -3)$ is the origin. With this hyperbola, $a^2 = 4$ and $c^2 = 9$. From this, we can find $b^2$:     $b^2 = c^2 - a^2$

$$b^2 = 9 - 4$$

$$b^2 = 5$$

The equation of the hyperbola is $\frac{y^2}{4} - \frac{x^2}{5} = 1$.

---

We will now turn our attention to the asymptotic behavior of hyperbolas. We have analyzed the asymptotic behavior of functions in previous chapters. Remember that rational functions can have vertical, horizontal, or oblique asymptotes. Hyperbolas have *oblique* asymptotes.

The asymptotic behavior of a function describes how the function behaves when either the dependent variable or the independent variable become very large in magnitude, or tend towards ∞. The asymptotic behavior of a hyperbola is very important in describing the graph of the hyperbola.

For large values of $x$ (and corresponding large values of $y$), a hyperbola behaves similar to a linear function. To describe this linear behavior, start with the equation of a hyperbola (the particular form does not matter), and simplify it under the assumption that $x$ and $y$ are much, much larger than 1. I will examine hyperbolas of the form $\frac{y^2}{a^2} - \frac{x^2}{b^2} = 1$. Isolate the term involving $y$ by adding $\frac{x^2}{b^2}$ to both sides: $\frac{y^2}{a^2} = \frac{x^2}{b^2} + 1$. If we assume that $x$ is very, very large, so that adding 1 to the term $\frac{x^2}{b^2}$ is hardly noticeable, we see that $\frac{y^2}{a^2} \sim \frac{x^2}{b^2}$. I am using the symbol $\sim$ to mean "behaves like." I cannot write that $\frac{y^2}{a^2}$ and $\frac{x^2}{b^2}$ are equal, because they are not. They are, however, very close. If $\frac{y^2}{a^2} \sim \frac{x^2}{b^2}$, then $y^2 \sim \frac{a^2 x^2}{b^2}$, and, taking the square root of both sides gives: $y \sim \pm \frac{ax}{b}$. The two lines, $y = \frac{ax}{b}$ and $y = -\frac{ax}{b}$ are called the **oblique asymptotes** of the hyperbola $\frac{y^2}{a^2} - \frac{x^2}{b^2} = 1$. Along the same reasoning, the oblique asymptotes of the hyperbola $\frac{y^2}{a^2} - \frac{x^2}{b^2} = 1$ are $y = \pm \frac{bx}{a}$.

These equations are not worth memorizing. If you understand the process of finding the asymptotes, you will avoid any confusion as to whether the slope of the oblique asymptotes are $\pm \frac{a}{b}$ or $\pm \frac{b}{a}$. Personally, I have difficulty memorizing formulas, and prefer to understand where the formulas come from. It may require an extra step or two when solving a problem, but those extra steps reinforce the concepts behind finding asymptotes. The key to finding the oblique asymptotes of a hyperbola is to ignore the constant in the standard equation of the hyperbola, and only focus on the terms involving $x$ and $y$.

## Example 3

Find the asymptotes of the following hyperbolas:

a. $\dfrac{y^2}{36} - \dfrac{x^2}{25} = 1$ 　　　　 b. $\dfrac{x^2}{121} - \dfrac{y^2}{144} = 1$

**Solution:** Isolate the term that involves $y$ and then assume that the constant 1 is negligible:

a. $\dfrac{y^2}{36} - \dfrac{x^2}{25} = 1$ : 　　　　 $\dfrac{y^2}{36} = \dfrac{x^2}{25} + 1$

$$\dfrac{y^2}{36} \sim \dfrac{x^2}{25}$$

$$y^2 \sim \dfrac{36x^2}{25}$$

$$y \sim \pm\dfrac{6x}{5}$$

The oblique asymptotes are: $y = \pm\dfrac{6x}{5}$ .

b. $\dfrac{x^2}{121} - \dfrac{y^2}{144} = 1$ : 　　　　 $\dfrac{x^2}{121} - \dfrac{y^2}{144} = 1$

$$\dfrac{y^2}{144} = \dfrac{x^2}{121} - 1$$

$$\dfrac{y^2}{144} \sim \dfrac{x^2}{121}$$

$$y^2 \sim \dfrac{144x^2}{121}$$

$$y \sim \pm\dfrac{12x}{11}$$

The oblique asymptotes are: $y = \pm\dfrac{12x}{11}$ .

The graph of a hyperbola can be translated horizontally or vertically. When a hyperbola is translated, every point on the hyperbola is translated by the same amount. In particular, the center, the foci, and the vertices of the hyperbola are translated. The equations $\frac{x^2}{a^2} - \frac{y^2}{b^2} = 1$ and $\frac{y^2}{a^2} - \frac{x^2}{b^2} = 1$ describe hyperbolas whose center is $(0, 0)$. We could write these equations as $\frac{(x-0)^2}{a^2} + \frac{(y-0)^2}{b^2} = 1$, and $\frac{(y-0)^2}{a^2} - \frac{(x-0)^2}{b^2} = 1$, if we wanted to explicitly include the location of the center of the hyperbola. In general, the equations of hyperbolas with centers located at $(h, k)$ can be written as:

$$\frac{(x-h)^2}{a^2} - \frac{(y-k)^2}{b^2} = 1 \text{ and } \frac{(y-k)^2}{a^2} - \frac{(x-h)^2}{b^2} = 1.$$

Be sure to associate the $x$-coordinate of the center with the term that involves $x$ in the equation of the hyperbola, and the $y$-coordinate of the center with the term that involves $y$ in the equation of the hyperbola. We can find $c$ by finding the distance between the center and one focus. The distance between the center and one of the vertices is $a$. The asymptotes will be translated by the same amount that the center is translated. The following table may help you organize this information and keep it all straight.

| Equation for the hyperbola | $\frac{(x-h)^2}{a^2} - \frac{(y-k)^2}{b^2} = 1$ | $\frac{(y-k)^2}{a^2} - \frac{(x-h)^2}{b^2} = 1$ |
|---|---|---|
| Transverse axis | Parallel to the $x$-axis | Parallel to the $y$-axis |
| Vertices | $(h \pm a, k)$ | $(h, k \pm a)$ |
| Foci | $(h \pm c, k)$ | $(h, k \pm c)$ |
| Asymptotes | $y - k = \pm\frac{b}{a}(x-h)$ | $y - k = \pm\frac{a}{b}(x-h)$ |

## Example 4

Find the equation of a hyperbola with center $(-3, 1)$, one focus at $(-3, 6)$, and one vertex at $(-3, 4)$.

**Solution:** The center, the focus, and the vertex all lie on the vertical line $x = -3$, so the transverse axis is parallel to the y-axis. We can therefore start with the equation $\dfrac{(y-k)^2}{a^2} - \dfrac{(x-h)^2}{b^2} = 1$ and start filling in the missing pieces. The center of the hyperbola is $(-3, 1)$, so $h = -3$ and $k = 1$. Our equation becomes $\dfrac{(y-1)^2}{a^2} - \dfrac{(x+3)^2}{b^2} = 1$.

To find $c$, we need to find the distance between the center and the focus. This distance is 5, so $c = 5$. To find $a$, we need to find the distance between the center and one vertex: the distance between $(-3, 1)$ and $(-3, 4)$ is $a = 3$. We can use the equation $b^2 = c^2 - a^2$ to find $b^2$: $b^2 = c^2 - a^2 = 5^2 - 3^2 = 25 - 9 = 16$.

The equation of the hyperbola is:

$$\frac{(y-1)^2}{9} - \frac{(x+3)^2}{16} = 1$$

In general, the equation of the oblique asymptotes is: $y - k = \pm\frac{a}{b}(x - h)$.

Substituting for the values of $a$, $b$, $h$, and $k$ gives: $y - 1 = \pm\frac{3}{4}(x + 3)$.

## Example 5

Find the vertices, the foci, and the oblique asymptotes of the hyperbola $x^2 - 3y^2 + 8x - 6y + 4 = 0$.

**Solution:** We need to complete the square in both $x$ and $y$ and put the resulting equation in standard form for a hyperbola:

$$x^2 - 3y^2 + 8x - 6y + 4 = 0.$$

Group the terms together $(x^2 + 8x) - (3y^2 + 6y) = -4$

Factor 3 from both terms involving $y$

$$(x^2 + 8x) - 3(y^2 + 2y) = -4$$

Complete the square $\quad (x^2 + 8x + 16 - 16) - 3(y^2 + 2y + 1 - 1) = -4$

Pull the unnecessary constant out of the parentheses. Make sure you multiply the constant by the coefficient in front

$$\left(x^2 +8x+16\right)-16-3\left(y^2 +2y+1\right)+3=-4$$

Factor the perfect squares $(x+4)^2 -16-3(y+1)^2 +3=-4$

Move the constants to the other side of the equation

$$(x+4)^2 -3(y+1)^2 = 9$$

Divide by 9 so that the constant term is 1

$$\frac{(x+4)^2}{9} - \frac{3(y+1)^2}{9} = 1$$

Cancel the 3s from the term involving $y$

$$\frac{(x+4)^2}{9} - \frac{(y+1)^2}{3} = 1$$

Now that the hyperbola is in standard form, we can read off the important information. The center is (−4, −1), $a^2 = 9$ and $b^2 = 3$. We can solve for $c$:        $b^2 = c^2 - a^2$

$$3 = c^2 - 9$$

$$c^2 = 12$$

$$c = \sqrt{12}$$

The transverse axis is parallel to the $x$-axis, the vertices are found using the expression $(h \pm a, k)$:

$$(h \pm a, k) = (-4 \pm 3, -1).$$

The vertices are (−7, −1) and (−1, −1). The foci are found using the expression $(h \pm c, k)$:  $(h \pm c, k) = (-4 \pm \sqrt{12}, -1)$.

The foci are $(-4 + \sqrt{12}, -1)$ and $(-4 - \sqrt{12}, -1)$. The oblique

asymptotes are found using the equation $y-k=\pm\frac{b}{a}(x-h)$:

$$y-k=\pm\frac{b}{a}(x-h)$$

$$y+1=\pm\frac{\sqrt{3}}{3}(x+4).$$

The oblique asymptotes are: $y=-1\pm\frac{\sqrt{3}}{3}(x+4)$.

There are many applications of hyperbolas in science and engineering. The properties of a hyperbola are used in radar tracking stations and navigation. An object can be located by sending sound waves from two sources. The sound waves from each source travel in concentric circles, and the intersection of the concentric circles from the two sources is hyperbolic. Hyperbolas are also used in astronomy. Although we typically think of the orbit of an object as an ellipse, the orbit of a comet is hyperbolic. Hyperbolas also appear when two quantities are inversely related to each other. For example, the volume and the pressure of a gas held at a fixed temperature are inversely related to each other, and their relationship is hyperbolic in nature.

## Lesson 10-3 Review

1. Find the coordinates of the foci and the vertices of the hyperbola described by $\dfrac{x^2}{25} - \dfrac{y^2}{49} = 1$. Find the equation of the oblique asymptotes and sketch its graph.

2. Find the equation of the hyperbola that has one vertex at $(3, 0)$ and foci at $(\pm 5, 0)$.

3. Find the equation of a hyperbola with center $(1, 4)$, one focus at $(-2, 4)$, and one vertex at $(0, 4)$.

4. Find the vertices, the foci, and the oblique asymptotes of the hyperbola $-x^2 + 2y^2 + 2x + 8y + 3 = 0$.

## Answer Key

### Lesson 10-1 Review

1. a. $y^2 = -10x: a = -10$,

   focus $\left(-\dfrac{10}{4}, 0\right)$,

   directrix: $x = \dfrac{10}{4}$,

   axis of symmetry is the $x$-axis,

   and the parabola opens upward.

   b. $x^2 = 4y: a = 4$,

   focus $(0, 1)$,

   directrix: $y = -1$,

   axis of symmetry is the $y$-axis,

   and the parabola opens to the left.

2. $x^2 = \frac{16}{3} y$

3. Axis of symmetry: $x = 3$,

    directrix: $y = -8$, $(x-3)^2 = 24(y+2)$.

4. Complete the square to find the equation for the standard form of the parabola: $(y+3)^2 = -1(x-9)$.

    $a = -1$,

    vertex: $(9, -3)$,

    focus: $\left(\frac{35}{4}, -3\right)$,

    directrix: $x = \frac{37}{4}$.

5. The receiver should be placed 0.64 feet from the base of the dish, along its axis of symmetry.

## Lesson 10-2 Review

1. The vertices are $(\pm 10, 0)$,

    the foci are $(\pm 6, 0)$,

    and the eccentricity is $e = \frac{c}{a} = \frac{6}{10} = \frac{3}{5}$.

    The graph of $\frac{x^2}{100} + \frac{y^2}{64} = 1$ is shown in Figure 10.19.

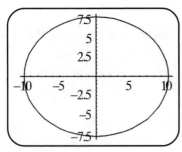

*Figure 10.19*

2. The equation of the ellipse is $\frac{x^2}{25} + \frac{y^2}{21} = 1$, and its eccentricity is $e = \frac{c}{a} = \frac{2}{5}$.

    The graph of $\frac{x^2}{25} + \frac{y^2}{21} = 1$ is shown in Figure 10.20 on page 223.

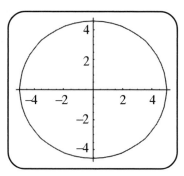

*Figure 10.20*

3. The equation of the ellipse is $\frac{x^2}{144} + \frac{y^2}{80} = 1$, and the foci are located at $(\pm 8, 0)$.

The graph of $\frac{x^2}{144} + \frac{y^2}{80} = 1$ is shown in Figure 10.21.

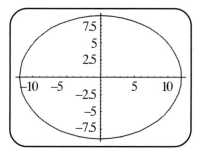

*Figure 10.21*

4. Major axis: $y = 3$,
   center: $(5, 3)$,

   the equation of the ellipse is $\frac{(x-5)^2}{49} + \frac{(y-3)^2}{24} = 1$.

## Lesson 10-3 Review

1. Foci: $\left( \pm\sqrt{74}, 0 \right)$,

   vertices: $(\pm 5, 0)$,

   oblique asymptotes: $y = \pm\frac{7x}{5}$.

   The graph of $\frac{x^2}{25} - \frac{y^2}{49} = 1$ is shown in Figure 10.22 on page 224.

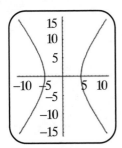

Figure 10.22

2. $\frac{x^2}{9} - \frac{y^2}{16} = 1$

3. $\frac{(x-1)^2}{1} - \frac{(y-4)^2}{8} = 1$

4. After completing the square, we have $\frac{(y+2)^2}{2} - \frac{(x-1)^2}{4} = 1$.

The center: $(1, -2)$, $a = \sqrt{2}$, $b = 2$, $c = \sqrt{6}$,

the vertices: $(1, -2 \pm \sqrt{2})$,

foci: $(1, -2 \pm \sqrt{6})$,

and the oblique asymptotes are $y + 2 = \pm \frac{\sqrt{2}}{2}(x - 1)$.

# Sequences and Series

It is natural to look for patterns around us. Patterns help us make sense of our world and enable us to make predictions and prepare for the future. Discovering weather patterns early enables us to prepare for a storm, and being familiar with traffic patterns can shorten our commute. Even though numbers are an abstract concept created by humans, there are numerical sequences, such as the Fibonacci sequence, appearing throughout nature.

## Lesson 11-1: Sequences

A **sequence** of numbers is an ordered list of numbers. The order in which the numbers appear is important. A sequence of numbers can be finite, such as the sequence 3, 5, 7, 11, 13; or the list can be infinite, such as the sequence 1, 3, 5, 7, 9, .... The first sequence is a list of the prime numbers less than 15, and the second sequence is a list of all of the odd numbers. The list of odd numbers is infinite because there is not a "last" odd number. We use three dots (...) to indicate that the list continues in the specified pattern.

The order in which a sequence of numbers is listed is important. For example, the sequence of numbers 3, 23, 6 may refer to the combination of a lock. If you try to enter the numbers in a different order, the lock will not open.

We can think of a sequence of numbers in terms of functions. A **finite sequence** of $n$ terms can be thought of as a function whose domain is the set $\{1, 2, 3, ..., n\}$. Remember that a function can be defined as a collection of ordered pairs. For example, the sequence 3, 23, 6 can be thought of as the function $\{(1, 3), (2, 23), (3, 6)\}$. The first coordinate of each point reveals the order of the numbers, and the second coordinate of each point indicates the actual number in the sequence. An infinite sequence is a function whose domain is the set of all positive integers.

We can define a sequence in several ways. We can hope that the pattern of the sequence is clear. Consider the sequence 2, 4, 6, 8, 10, 12, 14, .... Hopefully this sequence of numbers looks familiar: it is a list of the positive even integers, in order. Defining a sequence by hoping that the pattern is clear is a dangerous approach. You are counting on the fact that anyone who looks at the sequence can read your mind, and figure out of what pattern you were thinking. If you define a sequence in this way, do not be surprised if there are people who do not see the same pattern you do.

Another way to define a sequence is to give a formula for how to find the $n$th term in the sequence. For example, the sequence defined by the function $f(n) = n^2$ is the sequence 1, 4, 9, 16, 25, 36, 49, 64, .... In defining a sequence using a formula, we are explicitly describing how to find every term in the sequence. We do not have to hope that the pattern is clear. To find the 10th term in the sequence, we would evaluate $f(10) = 10^2 = 100$. This formula is sometimes written as $a_n = n^2$. When defining a sequence using a formula, we usually call the function $a_n$ rather than $f(n)$. We can find any term in the sequence by substituting into the given formula. This method of defining a sequence is very clear, but there are some sequences for which it is difficult, if not impossible, to find a formula that fits. For example, mathematicians have been unable to find a pattern in the sequence of prime numbers, and finding a formula to describe the sequence 10, 15, 16, 21, 26, 27, 32, 33, 38, 39, ... may be difficult. The pattern in that sequence is to add 5 to the first term, then add 1 to the second term, add 5 to the third term, add 1 to the next term, add 5, add 1, and so on. It is easier to describe the pattern than it is to define a formula for this sequence.

## Example 1

Write the first 5 terms of the following sequences:

a. $a_n = 2n$

b. $a_n = 2n - 1$

c. $a_n = (-1)^n$

**Solution:** Use the formulas to evaluate the terms of the sequence, starting with $n = 1$:

| n | $a_n = 2n$ | $a_n = 2n - 1$ | $a_n = (-1)^n$ |
|---|---|---|---|
| 1 | $a_1 = 2(1) = 2$ | $a_1 = 2(1) - 1 = 1$ | $a_1 = (-1)^1 = -1$ |
| 2 | $a_2 = 2(2) = 4$ | $a_2 = 2(2) - 1 = 3$ | $a_2 = (-1)^2 = 1$ |
| 3 | $a_3 = 2(3) = 6$ | $a_3 = 2(3) - 1 = 5$ | $a_3 = (-1)^3 = -1$ |
| 4 | $a_4 = 2(4) = 8$ | $a_4 = 2(4) - 1 = 7$ | $a_4 = (-1)^4 = 1$ |
| 5 | $a_5 = 2(5) = 10$ | $a_5 = 2(5) - 1 = 9$ | $a_5 = (-1)^5 = -1$ |

Finding a formula of a sequence can be tricky. You have to be sure, beyond a shadow of a doubt, that the pattern you observe is actually the one intended by whoever created the sequence.

### Example 2

Write a formula for the general term of the following sequences:

a. $1, \dfrac{1}{2}, \dfrac{1}{3}, \dfrac{1}{4}, \dfrac{1}{5}, \ldots$

b. $1, -2, 3, -4, 5, -6, \ldots$

**Solution:** Write out the terms and try to observe the pattern:

a. $a_1 = 1, a_2 = \dfrac{1}{2}, a_3 = \dfrac{1}{3}, a_4 = \dfrac{1}{4}, a_5 = \dfrac{1}{5}$. We can see that, in

general, $a_n = \dfrac{1}{n}$.

b. The alternating sign is a little tricky. It may help to string out the negative signs. Remember that the product of an even number of negative numbers will be a positive number, and the product of an odd number of negative numbers will be a negative number. $a_1 = 1, a_2 = -2, a_3 = 3 = -(-3), a_4 = -4 = -(-(-4))$, $a_5 = 5 = -(-(-(-5)))$. We can see that, in general, $a_n = (-1)^{n-1} n$. The term that produces the sign of the number in the sequence is not unique. We could also have defined the sequence using the formula $a_n = (-1)^{n+1} n$. Just make sure that the formula alternates sign, and starts out positive.

Another method that is used to define a sequence is called a recursive definition. A **recursive definition** of a sequence specifies how the sequence starts as well as how to find the *next* term in the sequence. For example, suppose we have a sequence that starts out 1, 1, 2, 3, 5, 8, 13, 21, 34, ... and the next term in the sequence is found by adding the previous two terms in the sequence together. This is a well-known sequence, called the Fibonacci sequence. The Fibonacci numbers are a very interesting collection of numbers. They can be used to model bee populations, and the sequence appears in a variety of unexpected places such as biology and architecture.

## Example 3

Find the first five terms of the infinite sequence in which $a_1 = 5$ and $a_n = 2a_{n-1} + 1$.

**Solution:** Use the value of $a_1$ to generate the next term in the sequence:

| $n$ | $a_n$ |
|-----|-------|
| 1 | $a_1 = 5$ |
| 2 | $a_2 = 2a_{2-1} + 1 = 2a_1 + 1 = 2(5) + 1 = 11$ |
| 3 | $a_3 = 2a_{3-1} + 1 = 2a_2 + 1 = 2(11) + 1 = 23$ |
| 4 | $a_3 = 2a_{4-1} + 1 = 2a_3 + 1 = 2(23) + 1 = 47$ |
| 5 | $a_5 = 2a_{5-1} + 1 = 2a_4 + 1 = 2(47) + 1 = 95$ |

If the formula of a sequence is known, we usually abbreviate it using brackets: $\{a_k\}_{k=1}^{n}$ is used to denote a finite sequence of terms $a_1, a_2, ..., a_n$, and $\{a_k\}_{k=1}^{\infty}$ is used to denote the infinite sequence $a_1, a_2, a_3 ....$ For example, the sequence $\{2n - 1\}_{n=1}^{10}$ denotes the first 10 odd positive integers, and $\{2n\}_{n=1}^{\infty}$ represents the sequence of even positive integers. It is important to realize that the letter used to index the sequence is a "dummy" variable (it doesn't matter which letter we use). We could write the sequence $a_1, a_2, a_3, ...$ as $\{a_k\}_{k=1}^{\infty}$ or $\{a_n\}_{n=1}^{\infty}$ or $\{a_t\}_{t=1}^{\infty}$, depending on your mood. All three representations are equivalent.

## Lesson 11-1 Review

1. Write the first 5 terms of the sequence $a_n = 2^n$.

2. Write a formula for the general term of the sequence 3, –6, 9, –12, 15, ….

3. Find the first five terms of the sequence in which $a_1 = 1$ and $a_n = 3a_{n-1} + 2^n$.

# Lesson 11-2: Series

A sequence is a list of numbers. If we add the numbers in the list, we create a *series*. Writing out the sum of a long list of numbers can quickly become cumbersome, so we will need to develop some notation to compress the equations.

The sequence of the first $n$ numbers is the sequence 1, 2, 3, 4, …, 100, and it can be abbreviated by $\{k\}_{k=1}^{100}$. The series formed by adding these 100 numbers together is the series $1 + 2 + 3 + … + 100$. Adding the first 100 numbers together may take awhile, but in the next two lessons we will discover some quick and easy ways to add the terms of some specific types of sequences together.

Writing the first 100 numbers out is a time-consuming task in and of itself, which is why we use the abbreviation $\{k\}_{k=1}^{100}$. Writing out the sum of the first 100 numbers is also a time-consuming task (and it doesn't even cover doing the actual addition), so it is in our best interest to develop some notation for series. The notation for series makes use of the Greek letter sigma (capital sigma, actually): $\Sigma$. Suppose we have a finite sequence of numbers $\{a_k\}_{k=1}^{n}$. We will write the sum $a_1 + a_2 + a_3 + … + a_n$ as $\sum_{k=1}^{n} a_k$.

In other words, $\sum_{k=1}^{n} a_k = a_1 + a_2 + a_3 + … + a_n$, or $\sum_{k=1}^{n} a_k$, means the same thing as $a_1 + a_2 + a_3 + … + a_n$. Although $\sum_{k=1}^{n} a_k$ looks pretty intimidating, you shouldn't let it have the upper hand. I'm sure that you would appreciate this shorthand notation after writing out a series that involves 1,000 terms, and I'm almost positive that you would rather use this notation than write out a series that involves infinitely many terms. When using this notation, it is important to see what each piece means. The Greek letter sigma represents the sum of the terms in the series. The term $a_k$ represents the formula of the sequence of numbers that we will be adding. The $k = 1$ at the bottom of the sigma indicates the starting value of $k$, and the $n$ at the top of the sigma indicates the stopping value of $k$. The variable $k$ scrolls from its starting point to its ending value, and each value of $k$ is substituted for $a_k$. The variable $k$ is called the **index of summation**. Let us practice working with this notation.

## Example 1

Write the expanded meaning of the following series:

a. $\displaystyle\sum_{k=1}^{5} a_k$  b. $\displaystyle\sum_{k=1}^{5} 2k$  c. $\displaystyle\sum_{k=3}^{6} k^2$

**Solution:**

a. Use the definition of the summation notation:

$$\sum_{k=1}^{5} a_k = a_1 + a_2 + a_3 + a_4 + a_5.$$

b. In this case, the we are given the formula for the sequence: $a_k = 2k$. We can expand this notation in two steps:

$$\sum_{k=1}^{5} 2k = \sum_{k=1}^{5} a_k = a_1 + a_2 + a_3 + a_4 + a_5, \text{ where } a_k = 2k.$$

We can then find $a_1, a_2, a_3, a_4, a_5$ using the formula $a_k = 2k$:
$a_1 = 2, a_2 = 4, a_3 = 6, a_4 = 8, a_5 = 10.$
We can then substitute these values into our expanded

expression: $\displaystyle\sum_{k=1}^{5} 2k = 2 + 4 + 6 + 8 + 10.$

Looking at this last equation, we can see that we evaluate $2k$ as $k$ scrolls from 1 to 5, and then add the terms.

c. With this problem, the starting value of $k$ is 3 (rather than 1), and its ending value is 6. Substitute each integer value of $k$ between 3

and 6 into the formula $a_k = k^2$: $\displaystyle\sum_{k=3}^{6} k^2 = 3^2 + 4^2 + 5^2 + 6^2.$

We have practiced using summation notation, but we haven't actually found any sums. In the next example, we will actually find the sums.

## Example 2

Find the following sums:

a. $\displaystyle\sum_{k=1}^{4} 2^k$     b. $\displaystyle\sum_{k=1}^{5}(2k+1)$     c. $\displaystyle\sum_{k=1}^{4}\frac{(-1)^k}{k}$

**Solution:**

a. $\displaystyle\sum_{k=1}^{4} 2^k = 2^1 + 2^2 + 2^3 + 2^4 = 2+4+8+16 = 30$

b. $\displaystyle\sum_{k=1}^{3}(2k+1) = (2\cdot 1+1)+(2\cdot 2+1)+(2\cdot 3+1) = 3+5+7 = 15$

c. $\displaystyle\sum_{k=1}^{4}\frac{(-1)^k}{k} = \frac{(-1)^1}{1}+\frac{(-1)^2}{2}+\frac{(-1)^3}{3}+\frac{(-1)^4}{4}$

$\displaystyle = -1+\frac{1}{2}-\frac{1}{3}+\frac{1}{4} = -\frac{7}{12}$

---

It may take a little practice to get used to this new notation, but you'll get the hang of it. You'll also want to practice breaking this summation notation into pieces. For example, $\displaystyle\sum_{k=1}^{n} a_k = \sum_{k=1}^{p} a_k + \sum_{k=p+1}^{n} a_k$. The term $\displaystyle\sum_{k=1}^{p} a_k$ means to add the terms in the sequence $\{a_k\}$ as $k$ scrolls between 1 and $p$, and the term $\displaystyle\sum_{k=p+1}^{n} a_k$ means to add the terms in the sequence $\{a_k\}$ as $k$ continues its scroll from $p+1$ to $n$. It's just a different grouping of the terms in the series:

$$\sum_{k=1}^{n} a_k = a_1 + a_2 + \ldots + a_p + a_{p+1} + \ldots + a_n = \left(a_1 + a_2 + \ldots + a_p\right) + \left(a_{p+1} + \ldots + a_n\right)$$

$$= \sum_{k=1}^{p} a_k + \sum_{k=p+1}^{n} a_k$$

For example,

$$\sum_{k=1}^{10} a_k = a_1 + a_2 + \ldots + a_{10} = \left( a_1 + a_2 + \ldots + a_6 \right) + \left( a_7 + a_8 + \ldots + a_{10} \right)$$

$$= \sum_{k=1}^{6} a_k + \sum_{k=7}^{10} a_k$$

Another common breakup of $\sum_{k=1}^{n} a_k$ involves just peeling off the last

term in the series: $\sum_{k=1}^{n} a_k = \sum_{k=1}^{n-1} a_k + a_n$.

For example, $\sum_{k=1}^{100} a_k = \sum_{k=1}^{99} a_k + a_{100}$ and $\sum_{k=1}^{20} a_k = \sum_{k=1}^{19} a_k + a_{20}$.

We can also change the index of summation. The key is to change the index everywhere in the series.

## Example 3

Change the index of summation from 4 to 1 in the series

$$\sum_{k=4}^{7} (3k - 1).$$

**Solution:** The series starts when $k = 4$ and ends when $k = 7$. Let us introduce a new letter, $j$, to serve as our new index of summation. When $k = 4$, we want $j = 1$; $k = 5$ corresponds to $j = 2$; and so on. The value of $k$ is 3 more than $j$, so the relationship between $j$ and $k$ can be written as $k = j + 3$.

To change the index of summation, we need to get rid of all

instances of $k$ in the series $\sum_{k=4}^{7} (3k - 1)$ by replacing it with $j + 3$:

$$\sum_{j=1}^{4} (3(j+3) - 1) = \sum_{j=1}^{4} (3j + 8).$$

We can verify this substitution by comparing the terms in each of the series:

$$\sum_{k=4}^{7}(3k-1)=(3\cdot4-1)+(3\cdot5-1)+(3\cdot6-1)+(3\cdot7-1)$$
$$=11+14+17+20$$

and

$$\sum_{j=1}^{4}(3j+8)=(3\cdot1+8)+(3\cdot2+8)+(3\cdot3+8)+(3\cdot4+8)$$
$$=11+14+17+20$$

The two series match up, term for term, so our substitution did not change the value of the series.

---

As we learned in the previous example, the starting value of a series can be changed, as long as we are careful to change the corresponding formula for the sequence of numbers.

One application of a series of numbers involves calculating the arithmetic mean of a sequence of numbers. The **arithmetic mean** of two numbers is commonly called the average of the two numbers. The process for finding the average of several numbers involves adding all of the numbers and dividing by the number of terms that we added. This process can be represented using summation notation. If $\{a_k\}_{k=1}^{n}$ is a sequence of $n$ numbers, then the arithmetic mean of these numbers, which we will denote $\bar{a}$, can be found using the formula

$$\bar{a}=\frac{\displaystyle\sum_{k=1}^{n}a_k}{n}=\frac{a_1+a_2+...+a_n}{n}=\frac{1}{n}\left(\sum_{k=1}^{n}a_k\right).$$

## Example 4

Find the arithmetic mean of the numbers 80, 90, 75, 68, 82.

**Solution:** $\bar{a}=\dfrac{80+90+75+68+82}{5}=\dfrac{395}{5}=79$

## Lesson 11-2 Review

1. Find the following sums:

   a. $\displaystyle\sum_{k=1}^{5} 3^k$

   b. $\displaystyle\sum_{k=1}^{4}(3k-2)$

   c. $\displaystyle\sum_{k=1}^{4}\frac{(-1)^k}{2k}$

2. Change the index of summation from 6 to 1 in the series

   $$\sum_{k=6}^{10}\left(10k-2^k\right).$$

3. Find the arithmetic mean of the numbers 27, 32, 15, 21, 24.

# Lesson 11-3: Arithmetic Sequences and Series

An **arithmetic sequence** is defined as a sequence in which there is a common difference between consecutive terms. In other words, each pair of consecutive terms in the sequence differs by the same constant. For example, in the sequence 1, 5, 9, 13, ... each pair of consecutive terms differs by 4: $a_2 - a_1 = 5 - 1 = 4$, $a_3 - a_2 = 9 - 5 = 4$, and $a_4 - a_3 = 13 - 9 = 4$. This is an example of an arithmetic sequence. The constant difference between consecutive terms is called the **common difference**, and is denoted by $d$. The common difference will be positive if the sequence is increasing; the common difference will be negative if the sequence is decreasing.

Arithmetic sequences are special in part because it is possible to find a formula for the $n$th term in an arithmetic sequence. If an arithmetic sequence starts out with $a_1$ and has a common difference $d$, then the formula to find the $n$th term in the sequence is:

$$a_n = a_1 + (n-1)d$$

## Example 1

Find the 10th term in the arithmetic sequence 7, 20, 33, ....

**Solution:** For this sequence, $a_1 = 7$ and the common difference is $d = 20 - 7 = 13$. Using the formula $a_n = a_1 + (n-1)d$ with $n = 10$, we have $a_{10} = 7 + (10-1)(13) = 124$.

## Example 2

Find the 10th term of the arithmetic sequence with $a_3 = 10$ and $a_7 = 22$.

**Solution:** We are not given the first term in the sequence, nor are we told what the common difference is. We can substitute our limited information into the formula

$a_n = a_1 + (n - 1)d$ and try to answer the question:

$a_3 = 10 = a_1 + (3 - 1)d = a_1 + 2d$

$a_7 = 22 = a_1 + (7 - 1)d = a_1 + 6d.$

We now have a system of two equations:

$$\begin{cases} a_1 + 2d = 10 \\ a_1 + 6d = 22 \end{cases}$$

and two unknowns ($a_1$ and $d$). We can solve this system using any of the techniques discussed in Chapters 8 and 9: $d = 3$ and $a_1 = 4$. Now that we know $a_1$ and $d$, we can find $a_{10}$: $a_{10} = 4 + (10 - 1)(3) = 31$.

Arithmetic sequences can also be defined recursively. For example, the arithmetic sequence defined by $a_1 = 6$ and $a_n = a_{n-1} - 2$ starts out at 6 and the common difference is $-2$. The common difference is the difference between any two consecutive terms, and it can be determined by subtracting $a_{n-1}$ from both sides of the equation $a_n = a_{n-1} - 2$. The common difference is $d = a_n - a_{n-1} = -2$.

If an arithmetic sequence is defined by a formula, then by inspection you should be able to determine the first term and the common difference. Consider the arithmetic sequence defined by $a_k = 3k + 6$. The first term in the sequence is $a_1 = 3 \cdot 1 + 6 = 9$. The second term in the sequence is $a_2 = 3 \cdot 2 + 6 = 12$. Notice that the common difference is $12 - 9 = 3$. This is the constant in front of $k$ in the formula of the arithmetic sequence. In fact, the function $f(k) = 3k + 6$ should look familiar: it is a linear function with slope 3. The slope of the linear function corresponds to the value of the common difference in an arithmetic sequence. The common difference of the arithmetic sequence defined by $a_k = -2k + 3$ is $-2$, and the common difference of the arithmetic sequence defined by $a_k = 9k - 100$ is 9.

We can also work with arithmetic *series*. An **arithmetic series** is the sum of the terms in an arithmetic sequence. Finding the sum of an arithmetic sequence does not have to involve the tedious process of adding all of the terms. There is a clever observation that can be made with arithmetic series.

Consider the arithmetic sequence defined by $a_1 = 5$ and common difference equal to 10. Let $S$ denote the sum of the first 10 terms of this sequence. Then $S = 5 + 15 + 25 + 35 + 45 + 55 + 65 + 75 + 85 + 95$. Rather than adding all of these numbers together, let us write this sum in reverse order: $S = 95 + 85 + 75 + 65 + 55 + 45 + 35 + 25 + 15 + 5$. Now, let's line up these two equations and then add the corresponding terms together:

$$
\begin{array}{l}
S = 5 + 15 + 25 + 35 + 45 + 55 + 65 + 75 + 85 + 95 \\
+\ S = 95 + 85 + 75 + 65 + 55 + 45 + 35 + 25 + 15 + 5 \\
\hline
2S = 100 + 100 + 100 + 100 + 100 + 100 + 100 + 100 + 100 + 100
\end{array}
$$

Notice that the sum on the right involves the same number (100) added to itself 10 times, which happens to be the number of terms in the sequence that we were trying to add together. As a result, we have

$2S = (10)(100)$, or $S = \frac{(10)(100)}{2} = 500$. This method will work in general. If

$S_n$ denotes the sum of the first $n$ terms of an arithmetic sequence $\{a_k\}_{k=1}^{n}$, then one formula for finding $S_n$ is:

$$ S_n = \frac{n}{2}(a_1 + a_n) $$

This formula requires us to know the first and the last term of the arithmetic sequence. Using the formula $a_n = a_1 + (n - 1)d$, we can derive an equivalent formula for $S_n$:

$$ S_n = \frac{n}{2}(2a_1 + (n-1)d) $$

### Example 3

Find the sum of the following arithmetic series:

a. $\displaystyle\sum_{k=1}^{15}(3k - 5)$

b. $\displaystyle\sum_{k=3}^{10}(-2k + 10)$

**Solution:**

a. The first term in the sequence is $(3 \cdot 1 - 5) = -2$ and the common difference is 3. Using the second formula for the sum, we see that:

$$S_{15} = \frac{15}{2}(2(-2) + (15-1)3) = \frac{15}{2}(38) = 285 \ .$$

b. Notice that the series starts with $k = 3$ rather than $k = 1$. There are two ways to approach this problem. We can adjust the index to start at 1, or we can fiddle with the series. I'll work this problem out both ways, so you can see two of the options you have for solving this type of problem. First, we can adjust the index to start at 1 using the substitution $k = j + 2$ :

$$\sum_{j=1}^{8}(-2(j+2)+10) = \sum_{j=1}^{8}(-2j+6) \ .$$

The first term in this series is $a_1 = 4$, and the common difference

is $-2$, so $S_8 = \frac{8}{2}(2 \cdot 4 + (8-1)(-2)) = \frac{8}{2}(-6) = -24$ .

The second method involves fiddling with the series. Notice

that $\sum_{k=3}^{10}(-2k+10) = \sum_{k=1}^{10}(-2k+10) - \sum_{k=1}^{2}(-2k+10)$ .

The series $\sum_{k=3}^{10}(-2k+10)$ starts with $k = 3$ and ends with $k = 10$,

and the series $\sum_{k=1}^{10}(-2k+10)$ starts with $k = 1$ and goes

through $k = 10$. This second series contains some extra terms that we have to subtract out, which is why we have to

subtract $\sum_{k=1}^{2}(-2k+10)$. The series $\sum_{k=1}^{2}(-2k+10)$ represents

the extra terms that have to be removed. We can find each of these series and subtract. For both series, $a_1 = 8$ and $d = -2$. The only difference between the two series is the number of terms in the sequence that are being added. The first series adds up to:

$$\sum_{k=1}^{10}(-2k+10)=S_{10}=\frac{10}{2}\big(2(8)+(10-1)(-2)\big)=\frac{10}{2}(-2)=-10,$$

and the second series adds up to:

$$\sum_{k=1}^{2}(-2k+10)=S_{2}=\frac{2}{2}\big(2(8)+(2-1)(-2)\big)=\frac{2}{2}(14)=14.$$

From this, we have

$$\sum_{k=3}^{10}(-2k+10)=\sum_{k=1}^{10}(-2k+10)-\sum_{k=1}^{2}(-2k+10)=-10-14=-24.$$

Both methods yield the same result. I recommend understanding both methods, because they each look at the series $\sum_{k=3}^{10}(-2k+10)$ from a different angle.

### Lesson 11-3 Review

1. Find the 10th term in the arithmetic sequence 96, 90, 84, ....

2. Find the 10th term of the arithmetic sequence with $a_3 = 15$ and $a_7 = 45$.

3. Find the sum of the following arithmetic series:

   a. $\sum_{k=1}^{20}(4k-10)$    b. $\sum_{k=5}^{15}(3k-8)$

## Lesson 11-4: Geometric Sequences and Series

An arithmetic sequence is a sequence with a constant difference between consecutive terms. We made use of that property and developed a formula for the sum of the first $n$ terms of an arithmetic sequence. A **geometric** sequence is a sequence in which the *ratio* between consecutive terms is a constant. We will use $r$ to represent the common ratio.

The sequence 1, 2, 4, 8, 16, ... is an example of a geometric sequence. The next term in the sequence is found by doubling the previous term, and the common ratio for this sequence is 2. The formula for finding the $n$th term in the sequence is $a_n = 2^{n-1}$. We can verify that $a_1 = 2^{1-1} = 2^0 = 1$, $a_2 = 2^{2-1} = 2^1 = 2$, $a_3 = 2^{3-1} = 2^2 = 4$, and so on. In general, the $n$th term of a geometric sequence with common ratio $r$ and first term $a_1$ will be given by the formula $a_n = a_1 r^{n-1}$. This is actually an exponential function. When working with geometric series, it is important to keep in mind all of the techniques we developed to work with exponential functions.

## Example 1

List the first four terms of the geometric sequence whose first term is 5 and the common ratio is $\dfrac{1}{2}$.

**Solution:** The $n$th term of this geometric sequence is given by the formula $a_n = 5\left(\dfrac{1}{2}\right)^{n-1}$ .

Using this formula, we have: $a_1 = 5\left(\dfrac{1}{2}\right)^{1-1} = 5$,

$$a_2 = 5\left(\frac{1}{2}\right)^{2-1} = 5\left(\frac{1}{2}\right)^{1} = \frac{5}{2}, \quad a_3 = 5\left(\frac{1}{2}\right)^{3-1} = 5\left(\frac{1}{2}\right)^{2} = \frac{5}{4}, \text{ and}$$

$$a_4 = 5\left(\frac{1}{2}\right)^{4-1} = 5\left(\frac{1}{2}\right)^{3} = \frac{5}{8}.$$

If we are given two terms in a geometric sequence, we can find a formula for the $n$th term in the sequence. The two terms in the sequence can be considered two points that the function $f(n) = a_1 r^{n-1}$ passes through. We practiced solving these types of problems in Chapter 7.

## Example 2

Find the formula for the $n$th term in the geometric sequence if $a_2 = 6$ and $a_5 = 48$.

**Solution:** Solving this problem is equivalent to solving the problem of finding the function $f(n) = a_1 r^{n-1}$ that passes through the point (2, 6) and (5, 48). To find this equation, we need to substitute these data points into the formula $f(n) = a_1 r^{n-1}$ and solve for $a_1$ and $r$: $f(2) = 6 = a_1 r^{2-1} = a_1 \cdot r$ and $f(5) = 48 = a_1 r^{5-1} = a_1 \cdot r^4$. Now that we have two equations, $a_1 \cdot r = 6$ and $a_1 \cdot r^4 = 48$, we can solve them simultaneously. Taking the ratio of the two equations, we have

$\dfrac{a_1 \cdot r^4}{a_1 \cdot r} = \dfrac{48}{6}$. The nice thing about working with exponential

functions is that there are plenty of opportunities to cancel common factors. We can also use our quotient rule for exponents (when you divide two numbers with the same base, you subtract the exponents):

$$\frac{\cancel{a_1} \cdot r^4}{\cancel{a_1} \cdot r} = \frac{48}{6}$$

$r^3 = 8$

$r = 2$

Now that we know the common ratio, we can find the first term in the sequence:

$a_1 \cdot r = 6$

$a_1 \cdot 2 = 6$

$a_1 = 3$

The general formula for this sequence is $a_n = 3 \cdot 2^{n-1}$.

---

The common ratio of a geometric sequence does not have to be a positive number. The common ratio of the geometric sequence 5, −10, 20, −40, 80, −160, ... is −2. If you are given two odd terms, or two even terms of a geometric sequence, there may actually be two geometric sequences that fit the specific conditions. Remember that when you solve for the base of an exponent by taking an even root, there are actually two solutions: $r^n = c$ has solutions $r = \pm\sqrt[n]{c}$ if $n$ is even. Each value of $r$ leads to its own corresponding geometric sequence.

Keep in mind that the formula for the $n$th term in a geometric sequence, $a_n = a_1 r^{n-1}$ actually involves four variables: $n$, $a_n$, $a_1$, and $r$. If you know three of these values, you can always find the value of the fourth. There are many similarities between arithmetic sequences and geometric sequences. For example, in both sequences, finding the formula for the $n$th term only requires you to know two terms in the sequence. An arithmetic sequence is related to linear functions, and geometric sequences are related to exponential functions.

When we added the terms in an arithmetic sequence, we created an arithmetic series. When we add the terms in a geometric sequence, we will create a geometric series. In the last lesson, I presented a formula for adding the first $n$ terms in an arithmetic series. There is also a formula for adding the first $n$ terms in a geometric series. The sum of the first $n$ terms, $S_n$, of a geometric sequence with first term $a_1$ and common ratio $r$ (with $r \neq 1$) can be found by the formula $S_n = \dfrac{a_1(1-r^n)}{1-r}$. This formula works for ratios that are positive or negative.

## Example 3

Find the sum of the first 5 terms of the geometric series with $a_1 = 3$ and $r = -2$.

**Solution:** Use the formula $S_n = \dfrac{a_1(1-r^n)}{1-r}$:

$$S_5 = \frac{3\left(1-(-2)^5\right)}{1-(-2)} = \frac{3(33)}{3} = 33.$$

## Example 4

Find the sum of the first 6 terms of the geometric series with $a_2 = 4$ and $a_4 = 1$ and $r > 0$.

**Solution:** We need to find $a_1$ and $r$. We are given two terms in the sequence, which we can use to find $a_1$ and $r$: $a_2 = 4 = a_1 r^{2-1} = a_1 \cdot r$, and $a_4 = 1 = a_1 r^{4-1} = a_1 \cdot r^3$. Combining these two equations, which simplify to give $a_1 \cdot r = 4$ and $a_1 \cdot r^3 = 1$, have $r = \dfrac{1}{2}$ and $a_1 = 8$. Now we can use the formula $S_n = \dfrac{a_1(1-r^n)}{1-r}$: $S_6 = \dfrac{8\left(1-\left(\frac{1}{2}\right)^6\right)}{1-\frac{1}{2}} = \dfrac{8\left(\frac{63}{64}\right)}{\frac{1}{2}} = \dfrac{63}{4}.$

## Example 5

Evaluate $\sum_{k=1}^{4} 4\left(\frac{1}{3}\right)^k$ .

**Solution:** This is a geometric series with $a_1 = \frac{4}{3}$ (evaluate $4\left(\frac{1}{3}\right)^k$

when $k = 1$) and $r = \frac{1}{3}$ .

We are adding the first four terms of this geometric sequence

together. Using the formula $S_n = \dfrac{a_1\left(1-r^n\right)}{1-r}$ , we have

$$S_4 = \frac{\frac{4}{3}\left(1-\left(\frac{1}{3}\right)^4\right)}{1-\frac{1}{3}} = \frac{\frac{4}{3}\left(\frac{80}{81}\right)}{\frac{2}{3}} = \frac{160}{81}.$$

---

If the absolute value of the common ratio of a geometric sequence is less than 1, it is possible to add infinitely many terms of the geometric sequence together. The concept of addition of real numbers was developed using a finite number of finite numbers. Adding infinitely many numbers together requires us to extend our notion of addition, and the rules for adding infinitely many numbers together are developed in calculus. For now, I will only focus on adding infinitely many terms of a geometric sequence whose common ratio is less than 1. If $|r| < 1$, then $r^n$ becomes very, very small as $n$ gets large. We say that $r^n \to 0$ as $n \to \infty$. As a result, when $n$ becomes large, $1 - r^n$ becomes very close to 1, and $\frac{a_1(1-r^n)}{1-r}$

will get closer and closer to $\frac{a_1}{1-r}$ . The sum of infinitely many terms of a geometric sequence with $r < 1$, which we will denote $S_\infty$, can be found

using the formula $S_\infty = \frac{a_1}{1-r}$ . We can write $\sum_{k=1}^{\infty} a_1 r^{k-1} = \frac{a_1}{1-r}$ . Infinite geo-

metric series can be used to convert repeating decimals into fractions. They can also be used to prove some strange things about sets.

## Example 6

Evaluate $\displaystyle\sum_{k=1}^{\infty} 3\left(\frac{1}{2}\right)^{k-1}$

**Solution:** Use the formula $\displaystyle\sum_{k=1}^{\infty} a_1 r^{k-1} = \frac{a_1}{1-r}$ : $\displaystyle\sum_{k=1}^{\infty} 3\left(\frac{1}{2}\right)^{k-1} = \frac{3}{1-\frac{1}{2}} = 6$

## Example 7

Write the following repeating decimals as fractions:

a. 0.33333...

b. 0.124124124124...

**Solution:**

a. We can think of 0.3333... as the sum

$3(0.1) + 3(0.01) + 3(0.001) + ...$, or $3(0.1) + 3(0.1)^2 + 3(0.1)^3 + ....$

This is a geometric series with $a_1 = 0.3 = \frac{3}{10}$ and common ratio

$0.1 = \frac{1}{10}$ .

The sum of this infinite series is:

$$S_{\infty} = \frac{\frac{3}{10}}{1-\frac{1}{10}} = \frac{\frac{3}{10}}{\frac{9}{10}} = \frac{3}{10} \cdot \frac{10}{9} = \frac{3}{9} = \frac{1}{3}.$$

b. We can think of 0.124124124124... as $\dfrac{124}{1000} + \dfrac{124}{(1000)^2} + \dfrac{124}{(1000)^3} + ....$ .

This is a geometric series with $a_1 = \dfrac{124}{1000}$ and $r = \dfrac{1}{1000}$ .

The infinite series is:

$$S_{\infty} = \frac{\frac{124}{1000}}{1-\frac{1}{1000}} = \frac{\frac{124}{1000}}{\frac{999}{1000}} = \frac{124}{1000} \cdot \frac{1000}{999} = \frac{124}{999}$$

After you convert a few repeating decimals to fractions, you may notice a pattern that will shorten the process of converting repeating decimals to fractions.

In the study of mathematics, observed patterns lead to conjectures, and systematic techniques for solving problems are developed and generalized. Mathematics builds on previous results, and the material discussed in this book forms the foundation on which calculus is built.

## Lesson 11-4 Review

1. List the first four terms of the geometric sequence whose first term is 9 and the common ratio is $\frac{1}{3}$.

2. Find the formula for the $n$th term in the geometric sequence if $a_2 = \frac{1}{20}$ and $a_7 = \frac{8}{5}$.

3. Find the sum of the first 5 terms of the geometric series with $a_1 = 10$ and $r = -\frac{1}{2}$.

4. Find the sum of the first 6 terms of the geometric series with $a_2 = 10$ and $a_5 = \frac{2}{25}$.

5. Evaluate $\sum_{k=1}^{5} 10\left(\frac{1}{2}\right)^k$.

6. Evaluate $\sum_{k=1}^{\infty} 5\left(\frac{1}{3}\right)^{k-1}$.

7. Write 0.575757... as a fraction.

## Answer Key

### Lesson 11-1 Review

1. $2, 4, 8, 16, 32$
2. $a_n = (-1)^{n-1}(3n)$
3. $1, 5, 23, 85, 287$

## Lesson 11-2 Review

1. a. $\sum_{k=1}^{5} 3^k = 363$    b. $\sum_{k=1}^{4} (3k-2) = 22$    c. $\sum_{k=1}^{4} \frac{(-1)^k}{2k} = -\frac{7}{24}$

2. Change the index of summation from 6 to 1 in the series

$$\sum_{k=6}^{10} \left(10k - 2^k\right) = \sum_{j=1}^{5} \left(10j + 50 - 32\left(2^j\right)\right).$$

3. $\bar{a} = \frac{119}{5}$

## Lesson 11-3 Review

1. $a_1 = 96, d = -6, a_{10} = 42$

2. $a_1 = 0, d = 7.5, a_{10} = 67.5$

3. a. $\sum_{k=1}^{20} (4k - 10) = 640$    b. $\sum_{k=5}^{15} (3k - 8) = 242$

## Lesson 11-4 Review

1. $9, 3, 1, \frac{1}{3}$

2. $a_n = \frac{1}{40} 2^{n-1}$

3. $S_5 = \dfrac{10\left(1 - \left(-\frac{1}{2}\right)^5\right)}{1 - \left(-\frac{1}{2}\right)} = 6.875$

4. $a_1 = 50, r = \frac{1}{5}, S_6 = \dfrac{50\left(1 - \left(\frac{1}{5}\right)^6\right)}{1 - \frac{1}{5}} = 62.496$

5. $\sum_{k=1}^{5} 10\left(\frac{1}{2}\right)^k = 9.6875$

6. $\sum_{k=1}^{\infty} 5\left(\frac{1}{3}\right)^{k-1} = \frac{15}{2}$

7. $\frac{19}{33}$

# Index

# About the Author

*DENISE SZECSEI* earned Bachelor of Science degrees in physics, chemistry, and mathematics from the University of Redlands, and she was greatly influenced by the educational environment cultivated through the Johnston Center for Integrative Studies. After graduating from the University of Redlands, she served as a technical instructor in the U.S. Navy. After completing her military service, she earned a Ph.D. in mathematics from the Florida State University. She has been teaching since 1985.

# HOMEWORK HELPERS™

## The Essential Help You Need When Your Textbooks Just Aren't Making the Grade!

**Great preparation for the SAT II and AP Courses**  CAREER PRESS